T0265130

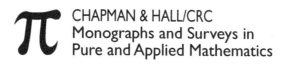

CHAPMAN & HALL/CRC
Monographs and Surveys in
Pure and Applied Mathematics   117

STRAIN SOLITONS

IN SOLIDS

and How to Construct Them

CHAPMAN & HALL/CRC
Monographs and Surveys in Pure and Applied Mathematics

Main Editors

H. Brezis, *Université de Paris*
R.G. Douglas, *Texas A&M University*
A. Jeffrey, *University of Newcastle upon Tyne (Founding Editor)*

Editorial Board

H. Amann, *University of Zürich*
R. Aris, *University of Minnesota*
G.I. Barenblatt, *University of Cambridge*
H. Begehr, *Freie Universität Berlin*
P. Bullen, *University of British Columbia*
R.J. Elliott, *University of Alberta*
R.P. Gilbert, *University of Delaware*
R. Glowinski, *University of Houston*
D. Jerison, *Massachusetts Institute of Technology*
K. Kirchgässner, *Universität Stuttgart*
B. Lawson, *State University of New York*
B. Moodie, *University of Alberta*
S. Mori, *Kyoto University*
L.E. Payne, *Cornell University*
D.B. Pearson, *University of Hull*
I. Raeburn, *University of Newcastle*
G.F. Roach, *University of Strathclyde*
I. Stakgold, *University of Delaware*
W.A. Strauss, *Brown University*
J. van der Hoek, *University of Adelaide*

CHAPMAN & HALL/CRC
Monographs and Surveys in
Pure and Applied Mathematics   117

STRAIN SOLITONS

IN SOLIDS

and How to Construct Them

ALEXANDER M. SAMSONOV

CRC Press
Taylor & Francis Group
Boca Raton London New York

CRC Press is an imprint of the
Taylor & Francis Group, an **informa** business

A CHAPMAN & HALL BOOK

First published 2001 by Chapman & Hall/CRC

Published 2019 by CRC Press
Taylor & Francis Group
6000 Broken Sound Parkway NW, Suite 300
Boca Raton, FL 33487-2742

© 2001 by Taylor & Francis Group, LLC
CRC Press is an imprint of Taylor & Francis Group, an Informa business

First issued in paperback 2019

No claim to original U.S. Government works

ISBN-13: 978-0-367-45540-8 (pbk)
ISBN-13: 978-0-8493-0684-6 (hbk)

This book contains information obtained from authentic and highly regarded sources. Reasonable efforts have been made to publish reliable data and information, but the author and publisher cannot assume responsibility for the validity of all materials or the consequences of their use. The authors and publishers have attempted to trace the copyright holders of all material reproduced in this publication and apologize to copyright holders if permission to publish in this form has not been obtained. If any copyright material has not been acknowledged please write and let us know so we may rectify in any future reprint.

Except as permitted under U.S. Copyright Law, no part of this book may be reprinted, reproduced, transmitted, or utilized in any form by any electronic, mechanical, or other means, now known or hereafter invented, including photocopying, microfilming, and recording, or in any information storage or retrieval system, without written permission from the publishers.

For permission to photocopy or use material electronically from this work, please access www. copyright.com (http://www.copyright.com/) or contact the Copyright Clearance Center, Inc. (CCC), 222 Rosewood Drive, Danvers, MA 01923, 978-750-8400. CCC is a not-for-profit organiza-tion that provides licenses and registration for a variety of users. For organizations that have been granted a photocopy license by the CCC, a separate system of payment has been arranged.

Trademark Notice: Product or corporate names may be trademarks or registered trademarks, and are used only for identification and explanation without intent to infringe.

Visit the Taylor & Francis Web site at
http://www.taylorandfrancis.com

and the CRC Press Web site at
http://www.crcpress.com

Library of Congress Card Number 00-046602

Library of Congress Cataloging-in-Publication Data

Samsonov, A.M. (Alexander M.)
   Strain solitons in solids and how to construct them / Alexander M. Samsonov.
      p. cm.
   Includes bibliographical references and index.
   ISBN 0-8493-0684-1
   1. Solitons. 2. Solid-state physics. I. Title.

QC174.26.W28 S26 2000
530.15'5353—dc21   00-046602

Contents

Preface

Calm is the master of move
Dao De Jing, 26
(Ancient Chinese philosophy)

The intriguing combination of the theories of physics and mechanics of solids, the lack of observation of elastic solitary waves in practice, and the possible application of general nonlinear wave theory to basic models of nonlinear mechanics and elasticity led to the studies on which this book is based. It was necessary to improve the existing theory, understand how to construct a powerful deformation pulse in a wave guide without plastic flow of material or fracture, and eventually propose a direct method of strain soliton generation, detection, and observation.

Solitary waves in shallow water have existed since water first covered the earth. The first documentation of the existence of shallow water waves appeared in 1834 when J. Scott Russell wrote one of the most cited papers about what later became known as soliton theory. Russell observed propagation of a solitary wave in the narrow Glasgow-Edinburgh canal that is still in use today. Many enthusiasts tried later to generate solitary water waves under natural conditions, but it wasn't until 1995 that participants in the Nonlinear Wave Symposium at the Heriot-Watt University in Edinburgh successfully construct such a wave in the canal.[1]

Russell noted in 1834 that solitary waves keep their shapes, move with constant velocity proportional to their amplitude, and recover completely after head-on collisions. In 1895, Korteweg and his pupil, de Vries, derived an equation describing shallow water waves. They proved that nonlinearity and dispersion achieved the balance that led to the generation of a stable wave which can propagate over long distances without changes of shape or velocity.

Kruskal and Zabusky coined the term soliton in 1965. They studied the waves in nonlinear chains, and applied the KdV equation together with periodic boundary and initial conditions to describe the recurrence problem. They found that solitary waves passed through each other without change of form and with only a small change in phase, i.e., the waves exhibit the particle-like behaviour.

The writings of Ablowitz, Calogero, Faddeev, Flashka, Kruskal, Lax, Miura, Novikov, Newell, Phillips, Scott, Zabusky, Zakharov and others about general nonlinearity, internal water waves, nerve pulse dynamics, ion-acoustic waves in plasma, and population dynamics explained much to experts in nonlinear science, but did not cover the great variety of nonlinear phenomena in existence.

---

[1] A wonderful collection of photographs of the event can be found at www.ma.hw.ac.uk/solitons

The shallow water soliton on the Scott Russell Aqueduct on the Union
Canal near Heriot-Watt University, Scotland, July 12th, 1995.

Despite many experiments in seismology and acoustics, little progress was made in studying the nonlinear stresses and strains present in solids. After the first attempts to base a theoretical description of a bulk soliton in a solid on the KdV equation model, it became clear that ultrasonic generation of waves and theoretical estimations of wave parameters were not appropriate. Long and bulk solitary deformation waves in elastic wave guides have been observable only on paper as late as the 1980s.

Short modulated (often called envelope) solitary waves governed by the nonlinear Schroedinger equation were widely used in work on the physics of solids and on nonlinear acoustics. Another soliton in a solid ball chain discovered by Frenkel and Kontorova (1938) became the standard for crystalline lattice models. Solitons in solids were found to be nonlinear, quasi-stationary localized strain waves propagating along an interface (internal solitary waves) or inside wave guides (density solitons). Modern development of wave theory led to thorough studies of nonlinear dynamics of condensed matter, particularly of nonlinear elastic solids. New and intriguing phenomena like solitary waves and breathers were discovered, but much work remained to be done.

A soliton propagates without change of shape in a uniform wave guide, but its shape will vary in the presence of inhomogeneities. Focusing may occur, and the amplitude of a soliton will increase while its width will simultaneously decrease. The localized area of plasticity may arise, resulting in fracture of a wave guide.

The study of the behaviour of such waves has become important for scientists assessing the durability of elastic materials and structures, developing nondestructive testing methods, and determining physical properties of conventional materials like brass and steel and new elastic materials, particularly polymers. Another possible application of strain nonlinear waves relates to their retention of shape properties despite the dependence of amplitude, velocity, and other parameters on the material properties of an elastic wave guide. This characteristic may be important in developing nondestructive tests for pipelines because bulk waves satisfy detection requirements more easily and effectively than surface strain waves. Furthermore, nonlinear elastic waves (e.g., solitary waves of strain) are of considerable importance in studying seismology, acoustics, and the mechanics of impact. Knowledge of the theory and experimental techniques can also apply to the physics of fracture, introscopy, investigation of sudden destruction, long distance energy transfer, vibro-impact treatments of hard materials, and other engineering problems.

The aim of this book is to present the general theory of wave propagation in nonlinear elastic solids, taking into account modern mathematical physics principles including the nonlinear theory of elasticity. Another goal is to demonstrate how basic principles of theory can provide successful experiments in physics. It introduces researchers and graduate students in mechanics and condensed matter physics to current nonlinear wave science and modelling of various physical effects in solids. I hope it will stimulate interest in studying nonlinear wave dynamics in dissipative and/or inhomogeneous wave guides. These areas of study are of great importance in the fields of theoretical and experimental nonlinear physics of solids, nonlinear mechanics, and several engineering disciplines.

Many people contributed to the writing of this book. I am pleased and happy to thank, first, my co-authors, co-workers, and friends, Drs. G.V. Dreiden, A.V. Porubov, I.V. Semenova, and E.V. Sokurinskaya for all their work and advice over the last 15 years. Their friendships and the guidance they provided from their unique areas of expertise contributed greatly to our successful studies. I am indebted to my esteemed friends and colleagues, Professors V. Babich, V. Babitsky, D. Crighton, J. Engelbrecht, D. Fusco, V. Golant, O. Konstantinov, G. Maugin, L. Ostrovsky, Y. Ostrovsky, V. Palmov, O. Rudenko, L. Slepyan, V. Zakharov,and N. Zvolinsky for providing valuable insight, astute criticism, encouragement, discussions, and invitations to seminars and conferences over the years. I also want to thank Professor Jeffrey, the first reader of this manuscript, for his thorough review and helpful advice.

I dedicate this book to my late parents. My work on solitons in solids could not have been done without their support, encouragement, and patience.

Introduction

Even recently, one of the most surprising things in the mechanics and physics of solids was why there was no individual mention of a solitary density wave or soliton – a concept of crucial importance to contemporary science. The wave propagation model seems to be universal for different branches of condensed matter physics, providing a good sign of the existence of long and bulk solitary waves in solids. No one observed this until 1988. The development of high resolution optical methods of wave detection allowed scientists to assume that the problem was also theoretical.

The solitary strain wave propagation problem has been widely studied in theory since the 1970's, see Nariboli (1970), Kunin (1975), Ostrovsky and Sutin (1977) as well as recent reviews by Engelbrecht (1983), Maugin (1994), and Samsonov (1994) to name only a few. In parallel with the general interest in long strain solitons as a new wave phenomenon in solids, one can recognize attempts to apply these waves to the third order elastic moduli measurements, to the non-destructive evaluation techniques, and to fracture and strain energy transition processes, etc.

When a powerful long strain wave propagates in a nonlinear elastic and bounded solid, the curving of a wave front can increase rapidly right up to the irreversible deformations appearance. This phenomenon, undesirable for the stability of solids, can be balanced with the dispersion of the wave inside a wave guide, having a small, finite but not infinitesimal cross section area that is in close correspondence with similar balance in the shallow water wave theory. However, the velocity of a longitudinal (density) wave in solids is much greater than in liquids and close to the shear wave velocity that may be among the explanations of former unsuccessful soliton's observation in experiments in solids.

The mathematical problem of nonlinear elastic guided wave propagation can be reduced to the coupled highly nonlinear partial differential equations, the p.d.e. A question arises on how to extract the most simple but informative model p.d.e. suitable to describe the long nonlinear strain wave, detectable using modern registration methods. One of the main differences in the physics of solids with fluid dynamics is that even in linear one-dimensional wave problems the existence of longitudinal and transversal motions with comparable velocities is to be taken into account. Tension is also a distinctive feature of elastic solids, therefore a sign of deformation should be important for the correct statement of the problem. Another difference to be considered is between the solitary and a weak (elastic) shock wave; the last will attenuate, while the soliton should propagate without losses.

Therefore, in order to initiate the physical experiment one has to study:

- simplest reductions of the p.d.e.;
- types of travelling waves;
- those solutions of the p.d.e., which provide detectable signals;
- influence of inhomogeneity, impurity and dissipation in a wave guide;
- results of numerical simulation as the cheapest experiment;
- possible artifacts;
  to clarify:
- type of nonlinear elastic material to be used;
- wave detection method;
- wave generation method;
- a method, applicable to the strain solitary wave observation;
  and also to determine:
- experimental setup for generation and detection;
- budget limitations and time limits.

Hence, the more complicated the problem under study, the more simple, transparent and robust the experimental approach should be.

The advantages provided by contemporary numerical simulation of nonlinear waves may not only lead to the successful experiments in physics of waves in solids but also result in new data of nonstationary wave propagation, in details of generation process, in prediction of the tentative limits for applications. Conversely, being very sensitive about possible artifacts, computational data are expected to be verified, as much as possible, by the genuine physical experiments. However, the main advantages of solitary strain (or bulk density) waves, i.e., the permanent wave shape despite dispersion in a wave guide and the ability to transfer an elastic energy for a long distance with no significant losses, may be of crucial importance for physics and technology.

Following Tvergaard (1997) in description of main scales in (micro)mechanics of solids, we shall mention the following scales for behaviour of materials :

i) the Burgers vector proportional to 10^{-10} m

Atomistic calculations (e.g., in metal physics) require a large amount of atoms, then one will meet computer limitations in 10^7 degrees of freedom, and a material volume that can be studied in this scale is extremely small from the viewpoint of nonlinear elasticity.

ii) the elastic cell size , i.e., the dislocation spacing of order $10^{-7} \div 10^{-6}$ m.

It is often used for models of inelastic deformations, however a number of dislocations limits strongly a size of a material volume under consideration.

iii) size of grains or inclusions etc. is of order $10^{-6} \div 10^{-4}$ m.

The main conclusion is that the continuum mechanics methods are very useful if an "elementary" volume size is, at least, 1000 times larger than the Burgers vector. The continuum limit of elasticity describes atomic lattice deformations, dislocation motions, void growth, microcrack formation, etc.

As an example, one can mention the ductile fracture, occuring due to nucleation and an increase of small void until coalescence and macrocrack formation happens, that involves large strains near the voids, ruptures and plasticity zone formation, etc. A natural question is what is the cause of the large strain? To study an internal (macro)structure of solids (in particular, alloys, ceramics, polymers and glasses) much attention is to be paid to purely elastic behaviour.

In this book, the main attention will be paid to the theory, simulation, generation and propagation of strain solitary wave in a nonlinearly elastic straight cylindrical rod under finite deformations, as the simplest guided wave dynamics problem. First, rigorous quantitative description of nonlinear wave in the three-dimensional continuum is complicated, and to discover a qualitatively new phenomenon is useful to reduce the space dimension of the problem. In the rod problem it may be done without restrictive simplifications. Second, the solitary wave is a result of balance of nonlinearity and dispersion, and in a rod the nonlinearity is due to finite deformations and elastic features of a material, whilst the dispersion is caused by the finite cross section area. Third, the rod is expected to be the most suitable for experimental study of the bulk solitary wave. Moreover, after experiments in solitons in rods, much progress was achieved in long nonlinear wave theory and observation in complex wave guides, plates and thin layers, that will also be considered.

Recently the nonlinear wave theory in rods was developed using the simplifications based on some physical hypotheses concerning the type of an elastic deformation, not on rigorous mathematical analysis and transformations of initial equations and boundary conditions. It was quite probable in this way that not every factor was taken into consideration in the problem's statement. Moreover, nonlinear elasticity theory equations are different in reference and current configurations, see Lurie (1980). It was seen recently that in small (ca. 0.001) elastic relative deformations any refinements will be small also and will not provide any qualitatively new phenomenon in deformation. It is quite common in static problems. However, in wave dynamics, a moment exists after which small adds will lead to big effects in wave propagation. Formally, a distinction of the nonlinear wave problem is in the fact that the definition of the functional type of the leading order equation of the asymptotic expansion arises in the next order problem as the compatibility condition (or the condition of absence of secular terms), in which the small disturbances will contribute.

The usage of the Hamilton principle as the main tool to derive the nonlinear equation governing the wave propagation in a rod resulted in a new Doubly Dispersive Equation (DDE) for the longitudinal bulk strain wave, containing two comparable dispersion terms caused by proximity of shears and longitudinal strains in solids. The next result was in proof of existence of allowed intervals of velocities, outside which no solitary wave can propagate.

It led to the conclusion that the compression solitons may be excited by an initial elastic pulse with transonic velocity – the crucial conclusion for further experimental observation of them.

However, there are no ideal rods, and will the soliton exist in, say, a tapered or uneven rod? It was studied thoroughly and resulted in the theory of soliton focusing in geometrically inhomogeneous wave guides. The solitary wave may be amplified in these wave guides, doubling its amplitude and velocity. The study of soliton propagation in different inhomogeneous wave guides was performed, as well as its attenuation or focusing in rods embedded in external elastic medium. It was necessary to develop the new approach to solve the corresponding DDE with dissipative terms, or, generally, the nonlinear equations with dispersion and dissipation. New, explicit and exact solutions were found, and the method was shown to be applicable to various nonlinear problems. Numerical simulation was used as the main approach to model the nonlinear wave excitation and its behaviour in non-stationary problems for complex wave guides with inclusions, etc.

Real physical experiments in soliton generation and observations were based on the theory developed. Additional problems were solved, namely, how to detect the soliton, what material the wave guide should be made of and which pulse may be transformed into a solitary wave without any irreversible deformations?

The simplest idea was to register the elastic waves by means of contact, e.g., by the piezoelectric sensors. However, one cannot follow the evolution of the wave in this case, and the sensor would influence the wave under study, due, in particular, to scattering. Moreover, in the pressure estimations the error level is close to 30-50%, and the bulk long density wave of small amplitude will be 'differentiated' by such a sensor in two separated pulses located in its front and back, i.e., two different waves will be registered. For all reasons, optical interferometry methods are preferable as contactless and very sensitive. We used the optical holography methods that provided us with two consequent holographic wave images, which keep the whole information of the object under study. Both wave pictures reconstructed from these holograms interfere, and one can see the changes in comparison with the stationary state. Hence both light waves are going through the same optical path and will be distorted by any optical disturbance identically. It means that in reconstruction these disturbances, if any, will be subtracted in the final picture, which will contain the undistorted information of the object.

The method proposed for soliton generation and observation seemed to be quite fruitful. We studied the physics of the density solitons in different rods and plates with various characteristics, as well as the reflection of solitons and focusing, and solitary waves in layers.

The question arises, why may we call these waves discovered as the long strain solitons or density solitons in solids? In part, the whole study was motivated by a simple question: if the theory may be so well grounded and developed, why nobody observed solitons in solids, except in virtual reality?

In this book we try to prove the existence of these intriguing nonlinear wave phenomena in solids.

List of symbols

A - strain amplitude

(A, B, C) - the elastic moduli by Landau

$\mathbf{C}$ - the Cauchy-Green finite deformation tensor

$c_0^2 = E/\rho$ - the linear longitudinal wave velocity in a rod

$c_1^2 = \mu/\rho$ - the linear shear wave velocity in a rod

cn, dn, sn - the Jacobi elliptic functions

∂W - the lateral surface

D - the volume extension

$D(u)$- the diffusion coefficient

E - the Young modulus

$\mathbf{e}_0$ - the linear deformation tensor

$\mathbf{F}_k$ - the external forces

$\mathbf{G}$ - the Cauchy-Green tensor of measure of deformation

$\mathcal{H}$ - the Hilbert transform

ΔK - fringe shift in a hologram

K - kinetic energy density

k - wave number

(l, m, n) - the third order elastic moduli introduced by Murnaghan

$\mathcal{L}(u, u_t, u_x, \ldots x, t)$ - the Lagrangian density per unit volume

$\mathbf{P}$ - the Piola-Kirchhoff stress tensor

$\wp$ - the Weierstrass elliptic function

o.d.e. - ordinary differential equation

p.d.e. - partial differential equation

$\mathbf{r}$ - the vector-radius

S - the action functional

S - cross section area

t - time

$\mathbf{U} = \{U_k\}$ - the displacement vector, defined in the Lagrangian co-ordinates

$\{x_k\}, \ k = 1, 2, 3$ - Cartesian co-ordinate system

ρ - the material density

$I_k = I_k(\mathbf{C})$ - invariants of the Cauchy-Green deformation tensor $\mathbf{C}$

$\mathbf{\nabla U}$ and $(\mathbf{\nabla U})^T$ - the vector-gradient and its transpose

$\beta = 3E + 2l(1 - 2\nu)^3 + 4m(1 + \nu)^2(1 - 2\nu) + 6n\nu^2$ - coefficient of nonlinearity

$\delta \widetilde{A}$ - the elementary work

$\delta \mathbf{U}$ - virtual displacement

$(\gamma_1, \gamma_2, \gamma_3, \gamma_4)$ - the 4th order elastic moduli

ε - a small parameter

Δ - the Laplacean

κ - the third order elastic modulus introduced by Mooney

μm - micrometers

ν - the Poisson coefficient

(ν_1, ν_2, ν_3) - the third order elastic moduli introduced by Lame

Λ - an elastic wave length, in particular, the strain solitary wave length

λ - the soliton width in numerical simulation

λ - the light length

(λ, μ) - the second order elastic moduli introduced by Lame

η - the viscocompressibility coefficient

Π - the volume density of potential energy

ω - frequency

Letter subscripts with displacements and strains denote differentiations.

Primes denote differentiations.

Bold font is used for vectors and tensors.

Chapter 1

Nonlinear waves in elastic solids

1.1 Basic definitions

There is no exhaustive definition of a wave, one the most common phenomena of nature, and for this reason we will adopt the one used by Truesdell and Noll (1965):

A wave is a state moving into another state with a finite velocity.

This definition is refined often as: *A wave is a state moving into another state with a finite velocity and transferring energy.*

Therefore, from the general viewpoint a wave is a moving state and at the same time a motion itself. The energy transfer is typical for wave motion, whilst no substance motion is assumed. Physical intuition allows one to pick out the wave motion as a multi-faceted and intriguing physical phenomenon quite definitely in various natural objects and to note the uselessness of unflexible and rigorous set of definitions. Evidently, it is necessary to enter into details for any kind of further study and to discover general features of and general ideas to describe wave propagation in solids. Considering motions of any bounded object, one can subdivide motions in two families, at least, depending on whether the object moves as a whole (note the pendulum oscillations), or if it remains in the vicinity of an average position in space and time, e.g., violin string vibrations. To underline the relativity and imperfections of definitions, we have to note that a motion can be attributed to both classes simultaneously, just like an ocean wave goes to shore, while an elementary water volume (or particle) oscillates in two orthogonal directions in the vicinity of an average state.

A solid as the substance is defined, as a rule, in several ways. Following Lurie (1980), one can describe a solid as *a material volume, having a definite shape and resisting every forces, that tend to change its shape.* Kunin (1975) proposed to define a solid as *a crystalline material composed of a lattice of atoms with some symmetry.* First definition is used in macroelasticity theory, the second is important for solids with microstructure (the momentum theory), however, both are not exhaustive. Indeed, the nonlinear elasticity of solids is based, still implicitly, on a model of elastic potential of atomic interactions, while the second definition seems not to be rigorously applicable to amorphous, porous or granular solids. However each of them has highly nonlinear elastic features.

Depending upon a kind of a condensed matter, the mechanical energy can be accumulated in a kinetic form, or a potential one, or both. The last case is typical for a massive solid resistive to both the deformation process and the motion itself. In these solids, an energy transfer is provided by the atom's motion near an equilibrium. Moreover, we will consider here wave propagation caused by an initial perturbation of a solid, whereas, in general, an excitation could be initiated also by means of so-called self-organisation process during a finite time interval.

Some types of waves can accompany acousto-optical, thermomechanical, biomechanical and other complex physical phenomena in solids. They cannot be analysed in the framework of solid mechanics only, but in the coupled field theory; however the corresponding well-posed problems will contain necessarily the mechanical statement. For further details we will refer to papers by Eringen and Maugin, 1990; Fusco and Jeffrey, 1991; Maugin, 1994. Strongly nonconservative systems will not be analysed here, while some attention will be paid to the weakly dissipative elastic interactions, that seem to be typical for elastic bodies imposed into an external medium.

We will consider the wave motion that can be characterized by an initial excitation as a source of a motion, and energy and momentum transfer, but not a mass transfer. Several restrictions are to be included into consideration before we will be able to describe even generally the object under study. Concerning elastic waves, it is quite common to interpret the general definition as the absence of any motion of an elastic medium as a whole, then any local disturbance that is moving to a neighbouring point during a finite time interval (not infinitesimal), i.e., with finite phase velocity.

The next step is to introduce an isotropic elastic medium and to define two types of waves. *Longitudinal* waves are those in which a particle moves along the direction of propagation, whereas the motion is perpendicular to it in *transverse* waves. Waves are assumed to be uncoupled in linear elasticity; therefore longitudinal waves are called as the extension, or the compression, or the dilatation waves, while transverse waves can be the shear, or the rotational, or the equivoluminal waves, etc. For solids with free surface (that means usually a surface between condensed matter and air, or an interface of two condensed media with remarkably different densities) one should take into consideration the possible existence of the Rayleigh waves, also, in which a particle motion is parallel to propagation direction and perpendicular to the interface. The last waves vanish with the distance from the interface exponentially. Again, one should mention the relativity of definitions. Advanced theories of elasticity do not exclude possible coupling or polarization effects (see, e.g., Bland, 1969, for details) and deal with waves in bounded elastic structures like rods, plates, shells, frames, multi-body structural elements, etc., to say nothing about elastic nonlinearity influence.

We will describe the basic terms of wave motion using the partial differential equations theory. A wave is characterized usually by its profile and velocity. Only finite velocity waves will be considered here, and we will often refer to the water surface waves as visual aids. The profile of a wave can be either smooth or discontinuous, and from the physical viewpoint it depends upon the scales for space and time (and, therefore, for velocity) used for the phenomenon description: we can mention here the crack propagation as the smooth and the finite speed motion of the strong discontinuity inside a solid, resulting in the free surface. Formally one can define a weak discontinuity as that appearing in the highest order derivative in corresponding differential operator, while the discontinuity in a low order derivative results to the strong discontinuity of a wave profile or to an impact type of wave. Obviously, it corresponds to boundary and initial conditions defining a wave propagation problem together with an equation.

In many cases the profile of a wave can be described in time t and space coordinate x separately; however there are progressive waves, often called *travelling waves*, propagating with a velocity V and having a profile depending on linear combination $(x \pm Vt)$. The last is usually called the phase variable. From the physical viewpoint these waves are of crucial importance due to comparatively simple experimental arrangements necessary to generate, detect and observe them.

1.1.1 Wave terminology

Following any textbook on wave dynamics, one should begin with the one-dimensional in-space wave propagation problem governed by the scalar hyperbolic p.d.e. written for a field variable $u(x,t)$ in the form

$$u_{tt} - V^2 u_{xx} = 0. \tag{1.1}$$

for the wave having constant velocity V. It has the well known general solution:

$$u = f_1(x - Vt) + f_2(x + Vt) \tag{1.2}$$

where both arbitrary functions f_1, f_2 , defined in $\mathcal{C}^2$, constitute the d'Alembert wave solution. It contains the only dependence of two linear combinations $(x \pm Vt)$ for any initial and boundary conditions, i.e., it represents the travelling (progressive) waves of constant size and shape, propagating in two opposite directions. Writing the solution (1.2) explicitly in an amplitude dependent form of a harmonic wave with k as a wave number and ω as a frequency,

$$u = A\cos[k(x \pm \omega t/k)], \tag{1.3}$$

one should identify the *phase velocity* equal to $V = \omega/k$. If it is constant for any k, waves are called non-dispersive, if not, then a relationship $V = V(k)$ is not trivial, and waves with different wave numbers propagate with different velocities, i.e., they disperse. The relationship

$$\omega = \Omega(k) \tag{1.4}$$

is called the *dispersive relation*. Then $d\omega/dk \neq 0$ and the second quantity, having the dimension of velocity, namely, the *group velocity* arises as

$$c_g = d\Omega/dk, \tag{1.5}$$

that is equal to the energy transfer velocity caused by wave motion; see, e.g., Lighthill (1965) for further details and Bland (1969) for useful examples in elasticity. Evidently, $V = \omega/k$ is not necessarily equal to c_g, moreover they may have opposite signs.

The dispersion relation demonstrates how various components of the Fourier decomposition of any initial wave propagate at different velocities, that lead to the variation in a wave shape.

Nonlinearity in any dynamical system provides a new feature of a wave velocity, namely, a possible dependence of it upon the field variable itself,

$$u_{tt} - V^2(u)u_{xx} = 0. \tag{1.6}$$

Therefore a solution continuously changes its profile up to a weak shock (discontinuity) formation, that starts to propagate with different, and usually bigger than V, velocity, see (Drazin and Johnson, 1989). For example, assuming $V(u) \equiv a + u$, a - const, one has a solution in a form $u = f_1[x - (a + u)t]$, evidently varying with x and t. Moreover, some waves can propagate with amplitude dependent velocity, and in order to generate such a wave with stable or even constant profile the nonlinearity should be balanced with an 'inverse' phenomenon taken into account, e.g., with dissipation, or dispersion, or with the cumulative (hereditary) features of a medium. Dissipation is usually defined as the corresponding exponential decay factor in a wave solution (1.3), and occurs if the odd order derivatives are included into consideration in the hyperbolic model (1.1). When $V = \omega/k = c_G$, both the medium and the wave propagation model are neither dispersive nor dissipative.

Wave profile can be either smooth, $u \in C^2$ for (1.6), or discontinuous, which is in its turn either weak or strong, depending on the highest order of a discontinuous derivative in a equation. The strong discontinuity, often called a shock, appears when this order is less than the highest order of a derivative of a field variable involved in the differential operator. The weak discontinuities are referred to as the acceleration waves; see (Truesdell and Noll, 1965). Moreover, profiles can be of an impact type, of oscillating type, or represent the solitary wave, first observed by Russel, (1834) in shallow water and found by Korteweg and de Vries, (1895) as the solution to the corresponding new nonlinear equation, derived by themselves and are become the most important one in the nonlinear wave propagation theory. The solitary wave occurs as a result of a balance between nonlinearity and dispersion (and/or dissipation) and it will be of main interest throughout the book. Here we just mention some different profiles such as the bell-shaped solitons, the kink-shaped solitons (see, e.g., Drazin and Johnson, 1989; Dodd et al, 1982; Bullough and Caudrey, 1980, to name only a few) and even the asymmetric solitary waves (Engelbrecht, 1991).

Considering the simplest linear Klein-Gordon equation for an unknown $\phi(x,t)$:

$$\phi_{tt} - \phi_{xx} = m\phi \tag{1.7}$$

one has a dispersion relation

$$\omega = \omega(k) = \pm\sqrt{m + k^2} \tag{1.8}$$

The phase velocity defined as $c_{ph} = \omega/k$ describes how a surface of constant phase moves, while the group velocity $c_G = \partial\omega/\partial k$, describes how fast the bulk of the wave propagates. The word "dispersion" means that for real ω linear waves of different k will have different c_{ph} and c_G, and components of the wave will spread or disperse

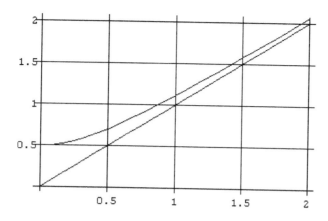

Figure 1.1: For $k \to 0$ $\omega(k)$ tends to $\sqrt{m}$ not to zero for the linear Klein-Gordon equation.

during propagation. This means that c_{ph} and c_G are $k-$dependent for $m \neq 0$,while for $m = 0$, this is not so in (1.7).

Another useful example of the (nonlinear) wave equation is the dimensionless non-linear double dispersive equation (DDE):

$$u_{tt} - u_{xx} = (u^2 + au_{tt} + bu_{xx})_{xx}; a, b - const, \qquad (1.9)$$

that will be derived below and is, to some extent, close to the first Boussinesq equation:

$$u_{tt} - u_{xx} = (u^2 + au_{tt})_{xx}; \qquad (1.10)$$

and to the nonlinear string or the second Boussinesq equation (as a rule, $b < 0$):

$$u_{tt} - u_{xx} = (u^2 + bu_{xx})_{xx}. \qquad (1.11)$$

Evidently, the dispersion relations determined as $\omega = \omega(k)$ for $u \propto \exp i(kx - \omega t)$ are different for these equations, as it is shown in Figure (1.2): Note that the dispersive relation analysis is applicable for small k only; that is a serious limitation of the approach.

The phase velocity for (1.9) is

$$c_{ph} = \sqrt{\frac{1 - bk^2}{1 + ak^2}} \qquad (1.12)$$

and the group velocity may be written as:

$$c_G = \frac{1 - 2bk^2 - abk^4}{(1 + ak^2)c_{ph}} \qquad (1.13)$$

For different values of a and b one can easily calculate both velocities for the double dispersive equation and for both Boussinesq equations; see Figure (1.3). Finally, the group velocities are shown for the same equations in Figure (1.4). Note the negative

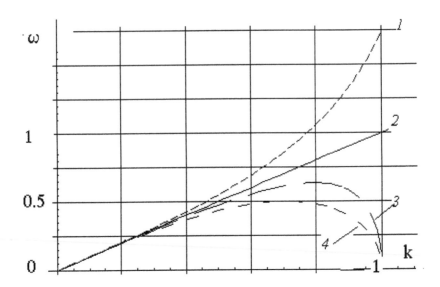

Figure 1.2: Dispersion relations for: *1*)-1st Boussinesq equation $b = 0, a = -2/3$; *2*) linear eq.; *3*) 2nd Boussinesq equation, $a = 0, b = 1$ and *4*) DDE with $a = -2/3, b = 1$.

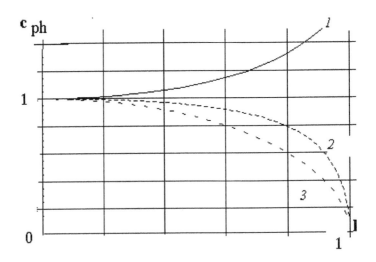

Figure 1.3: Phase velocities for: 1) 1st Boussinesq eq. with $a = -2/3, b = 0$; 2) DDE with $a = -2/3, b = 1$; 3) 2nd Boussinesq eq. with $a = 0, b = 1$.

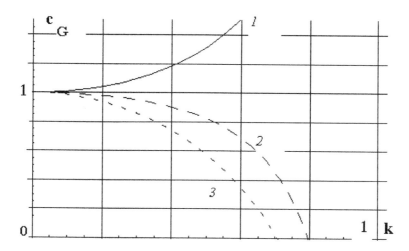

Figure 1.4: Group velocities for 1) 1st Boussinesq eq.; 2) DDE; and 3) 2nd Boussinesq eq. with the same values a and b, as above.

values of c_G even for $k < 1$; it means that, in general, waves with negative energy may propagate in the system considered; see, e.g., (Ostrovsky, Potapov, 1999). Some of these waves could propagate without energy transfer:

$$c_G = \partial\omega/\partial k = 0 \quad \text{for} \quad c_{ph} = \omega/k = \sqrt{(a-b)/(1+ak^2)} \neq 0.$$

Therefore the wave energy may be 'locked' or, probably, go back in a hyperelastic solid while the nonlinear wave itself propagates in positive direction of x.

We shall show that the DDE has a solitary wave bell-shaped solution in the form:

$$u = \frac{A}{2} \cosh^{-2} \sqrt{\frac{A}{A(1+a)+b}} \left(x \pm t\sqrt{1+A} \right), \tag{1.14}$$

and similar solutions for the Boussinesq equations may be found from it. Zakharov (1973) found the Lax pair and the wave collapse for the nonlinear string equation ($a = 0, b < 0$).

A brief description of the main features of waves and wave terminology above is not complete. Moreover, the exhaustive list of references related to the topic would be long enough, therefore further consideration will have to be more specific.

1.1.2 Deformation and strains

To consider processes in deformable solids, we begin with the postulates of existence of a continuum and their initial state. This state will change with time due to loading forces, therefore to define the relationships between *initial* and *current* (or actual, as it is often referred to) configuration of a continuum, one should introduce deformation as a process and strains as its result. For a solid it means the variation of a volume and a shape due to loading. We will follow the Lagrangian formalism, avoiding any kinematic description and concentrating on phenomena occurring in time inside solids, neglecting a rigid body motion of themselves. Therefore, one has to describe a deformation process or, in particular, a wave motion, in terms of initial coordinates of each particle of continuum. A macroscopic description of an elastic continuum will be used as well as the postulate concerning the deformation as a reversible process.

Let us introduce briefly some fundamental concepts of tensor calculus in application to the continuum mechanics. The location of a particle in an initial configuration, v, defined in the 3-dimensional space by the triad $\{x_k\}, k = 1, 2, 3$, is described by a vector-radius $\mathbf{r}(x_1, x_2, x_3)$. The current configuration, V, in time $t > 0$ is defined by another vector radius $\mathbf{R}(x_1, x_2, x_3; t)$ and resulted from a continuum's motion, or the displacement of particles, that allows us to introduce the displacement vector $\mathbf{u} = \mathbf{R} - \mathbf{r}$. The following formal introduction is aimed in part to demonstrate that one of the main sources of nonlinearity is in intrinsic features of deformation process, no matter in which coordinates it is described. We will concentrate mainly on quantities defined in the initial state v due to the relative convenience of deformation description and strains.

The vector basis is a set of three non-coplanar vectors $\mathbf{r}_s$, and the orthonormal triad $\mathbf{i}_s$ is the vector basis, such as $\mathbf{i}_s \cdot \mathbf{i}_s = 1$ for $s = k$, else $\mathbf{i}_s \cdot \mathbf{i}_s = 0$. Mutual vector basis is defined as the triad $\mathbf{r}^s$ with components, depending on a volume w of a parallelepiped, built on three basic vectors $\mathbf{r}_s$:

$$\mathbf{r}^1 = (\mathbf{r}_2 \times \mathbf{r}_3)/w, \ \ \mathbf{r}^2 = (\mathbf{r}_3 \times \mathbf{r}_1)/w, \ \ \mathbf{r}^3 = (\mathbf{r}_1 \times \mathbf{r}_2)/w, \ \ w = \mathbf{r}_1.(\mathbf{r}_2 \times \mathbf{r}_3).$$

Two scalar products g_{sk} and g^{sk} for vector bases in v, the initial $\mathbf{r}_s$ and the mutual one, $\mathbf{r}^s$, can be introduced by the vector triads:

$$\mathbf{r}_s \cdot \mathbf{r}_k = g_{sk}, \ \ \mathbf{r}^s \cdot \mathbf{r}^k = g^{sk}, \ \ \mathbf{r}^s \cdot \mathbf{r}_k = \delta_k^s = \{0, \ s \neq k; 1, \ s = k\}.$$

Then $\mathbf{r}_s = \partial \mathbf{r}/\partial x_s, \mathbf{r}^s = g^{sk}\mathbf{r}_k$ and consequently $\mathbf{R}_s = \partial \mathbf{R}/\partial x_s, \mathbf{R}^s = G^{sk}\mathbf{R}_k$, in the current configuration. Obviously, there is no difference between two vector bases for the orthonormal triads.

Consequently, one can introduce the deformation gradient as a tensor $\boldsymbol{\nabla}_0 \mathbf{R}$ defined in the *current* configuration through the vector gradient operator $\boldsymbol{\nabla}_0$, defined in the *initial* configuration,

$$\boldsymbol{\nabla}_0 = \mathbf{r}^s \frac{\partial}{\partial q^s}, \ \ \boldsymbol{\nabla}_0 \mathbf{a} = \mathbf{r}^s \frac{\partial \mathbf{a}}{\partial q^s} = \mathbf{r}^s \cdot \mathbf{r}^k \nabla_s^0 a_k, \tag{1.15}$$

that is applied to the basis $\mathbf{R}$ in the current configuration. Conversely, the tensor $\boldsymbol{\nabla}\mathbf{r}$, defined in an *initial* configuration, is a result of an application of the vector gradient (defined in current configuration) to a vector basis in the initial state[1]. Both tensors can be described as the inverse to each other:

$$\boldsymbol{\nabla}_0 \mathbf{R} = (\boldsymbol{\nabla}\mathbf{r})^{-1}, \ \ \boldsymbol{\nabla}\mathbf{r} = (\boldsymbol{\nabla}_0 \mathbf{R})^{-1} \tag{1.16}$$

---

[1] We are aiming to reduce the number of indices in the final statement of a problem in the initial state.

while, obviously, for the vector basis and the vector gradient, both defined in the same configuration, we obtain the tensor unit:

$$\nabla \mathbf{R} = \nabla_0 \mathbf{r} = \mathbf{E}.$$

Differentials will be expressed as follows:

$$d\mathbf{R} = d\mathbf{r} \cdot \nabla_0 \mathbf{R}, \ d\mathbf{r} = d\mathbf{R} \cdot \nabla \mathbf{r},$$

Both vector gradients can be transposed. Marking them with an upper index T we introduce the deformation measure tensors $\mathbf{G}$ and $\mathbf{g}$ as follows:

$$\nabla_0 \mathbf{R} \cdot (\nabla_0 \mathbf{R})^T = \mathbf{G}, \ \nabla \mathbf{r} \cdot (\nabla \mathbf{r})^T = \mathbf{g}.$$

The first one is called as the Cauchy-Green deformation *measure tensor* and defined in an *initial* basis, the second is named after Almansi and defined in the current basis. In no way are they equal to the tensor unit $\mathbf{E}$, except the case of a rigid body motion.

Applying (1.16) to the displacement vector $\mathbf{u}$, we obtain

$$\nabla_0 \mathbf{R} = \nabla_0 \mathbf{r} + \nabla_0 \mathbf{u} = \mathbf{E} + \nabla_0 \mathbf{u}$$

then

$$\mathbf{G} = (\mathbf{E} + \nabla_0 \mathbf{u})(\mathbf{E} + \nabla_0 \mathbf{u}^T) = \mathbf{E} + 2\mathbf{e}_0 + \nabla_0 \mathbf{u} \cdot \nabla_0 \mathbf{u}^T = \mathbf{E} + 2\mathbf{C}, \qquad (1.17)$$

where $\mathbf{e}_0 = (\nabla_0 \mathbf{u} + \nabla_0 \mathbf{u}^T)/2$ is the linear deformation tensor, while $\mathbf{C}$ is the Cauchy-Green deformation tensor, defined in an initial configuration. Finally we have for $\mathbf{C}$:

$$\mathbf{C} = \frac{1}{2}(\nabla_0 \mathbf{u} + \nabla_0 \mathbf{u}^T + \nabla_0 \mathbf{u} \cdot \nabla_0 \mathbf{u}^T). \qquad (1.18)$$

Its invariants (the numbers) are determined as usual:

$$I_1(\mathbf{C}) = tr \ \mathbf{C}, \quad I_2(\mathbf{C}) = (1/2)[(tr \ \mathbf{C})^2 - tr \ \mathbf{C}^2], \quad I_3(\mathbf{C}) = \det \mathbf{C} \qquad (1.19)$$

and can be expressed in terms of invariants of the measure tensor $\mathbf{G}$ as

$$I_1(\mathbf{C}) = (1/2)[I_1(\mathbf{G})-3], \quad I_2(\mathbf{C}) = (1/4)[I_2(\mathbf{G})-2I_1(\mathbf{G})+3],$$

$$I_3(\mathbf{C}) = (1/8)[I_3(\mathbf{G}) - I_2(\mathbf{G}) + I_1(\mathbf{G}) - 1].$$

It is important to express some quantities, having transparent physical interpretation, in the introduced terms. In particular, in Cartesian coordinates x_i the components of $\mathbf{C}$ can be expressed as follows:

$$C_{ik} = e_{0,ik} + \frac{1}{2} \frac{\partial u_j}{\partial x^i} \frac{\partial u^j}{\partial x^k}$$

where $e_{0,ik}$ are components of the linear deformation tensor. It can be shown that the main relative extension is equal to $\delta_k \equiv (dS_k - ds_k)/ds_k = \sqrt{G_k}$. The volume extension D is a relationship of a difference in elementary volumes $(dV - dv)$ to the initial elementary volume dv, and it equals to

$$D = (dV - dv)/dv = \sqrt{I_3(\mathbf{G})} - 1 = \sqrt{1 + 2I_1(\mathbf{C}) + 4I_2(\mathbf{C}) + 8I_3(\mathbf{C})} - 1$$

This equality can be expressed in terms of three main components C_k of $\mathbf{C}$ as follows:

$$D = \sqrt{(1 + 2C_1)(1 + 2C_2)(1 + 2C_3)} - 1$$

that is, in each volume element the deformation can be considered as an aggregate of three independent deformations along three orthogonal directions-three main axes of tensor $\mathbf{C}$. Note that a reduction of a tensor to main axes in any prescribed point does not necessary mean that it is diagonal in other points.

1.1.3   Stresses

Let us assume that in an undeformed body all molecules are in thermal equilibrium, and all parts are in mechanical balance. Being deformed, a solid becomes unbalanced, that results in internal forces, which are mentioned as internal stresses, caused by internal molecular interactions. The radius of action is assumed to be very small for these stresses in classical elasticity, that is, free from any electrical or other coupled field consideration.

Let us define for the initial time moment $t = 0$ the Lagrangian coordinates $\{x_k\}$, for material particles of an elastic medium, that occupies a volume v and has the density ρ_0. When for $t > 0$ this coordinate system will be deformed, following the medium particles, it provides variations of volume to V and density - to ρ, respectively.

The following conservation laws will have to be postulated as the cornerstones of the theory in mechanics of solids. To derive any governing equation one should take into account:

i) conservation of mass:

$$\int_v \rho_0 dv = \int_V \rho dV \tag{1.20}$$

then in differential form,

$$\rho_0/\rho = dV/dv = J(\mathbf{r}; \mathbf{R}) \tag{1.21}$$

that is equal to the Jacobian of a medium transformation from the initial configuration $\mathbf{r}$ to the current one, $\mathbf{R}$. In other words, the deformation is one-valued, and the differential form (1.21) of mass conservation can be called the continuity equation. Moreover, to introduce stresses, one should formulate:

ii) the balance of momentum in the form:

$$\nabla \cdot \mathbf{T} + \rho_0(\mathbf{k} - \mathbf{a}) = 0, \tag{1.22}$$

where $\mathbf{k}$ is a bulk force per mass unit and $\mathbf{a} = d\mathbf{v}/dt$ is the acceleration vector, ∇ is the deformation gradient (i.e., the tensor) multiplied to the Cauchy stress tensor $\mathbf{T}$, defined in *current* configuration. The significance of introduction of $\mathbf{T}$ is that the stress vector $\mathbf{t}$ on a small element of area, having a normal $\mathbf{N}$, can be found by means of the Cauchy formula:

$$\mathbf{t} = \mathbf{N} \cdot \mathbf{T}.$$

The tensor $\mathbf{T}$ is symmetrical, and the normal stress σ_N across an elementary area $\mathbf{N}dS$ is defined as the scalar product $\sigma_N = \mathbf{t} \cdot \mathbf{N} = \mathbf{N} \cdot \mathbf{T} \cdot \mathbf{N}$. In the Cartesian coordinates the physical stresses are identical with the components of $\mathbf{T}$.

The formal problem of the stress determination is that the actual configuration appears as a result of deformation process and cannot be determined before loading. To solve it, one can introduce the Piola-Kirchhoff stress tensor $\mathbf{P}$ (or the Lagrangian stress tensor) defined in the *initial* configuration that can be evaluated in terms of the vector gradient transpose $\boldsymbol{\nabla}\mathbf{r}^T$, and the square root of the determinant of the Cauchy-Green deformation measure tensor $\mathbf{G}$, defined in (1.17),

$$\mathbf{P} = \sqrt{G/g}\, \boldsymbol{\nabla}\mathbf{r} \cdot \mathbf{T}, \tag{1.23}$$

because from (1.21) one has $\det \mathbf{G} = G/g = (dV/dv)^2$. The balance of moment of momentum in local form for non-polar material leads to the symmetry of the stress tensor $\mathbf{P}$ components. We will not consider the so-called polar media, in which at any surface there is also moment stress vector in addition to the stress vector. Details, which are necessary to derive the polar medium equation, can be found elsewhere (Bland, 1969; Lurie, 1980; Maugin, 1988).

An elementary work $\widetilde{\delta}A$ (not a variation!) of external both mass and surface forces on a *virtual* displacement $\delta\mathbf{R}$ of particles from an initial to a current configuration of a volume V is defined as:

$$\widetilde{\delta}A = \int\!\!\!\int\!\!\!\int_V \rho\mathbf{k} \cdot \delta\mathbf{R}dV + \int\!\!\!\int_S \mathbf{f} \cdot \delta\mathbf{R}dS. \tag{1.24}$$

Transformation of integrals to those defined in the initial configuration v, with (1.23) being taken into account, yields

$$\widetilde{\delta}A = \int\!\!\!\int\!\!\!\int_v \rho_0\mathbf{k} \cdot \delta\mathbf{R}dv + \int\!\!\!\int_s \mathbf{n} \cdot \mathbf{P} \cdot \delta\mathbf{R}ds = \tag{1.25}$$

$$= \int\!\!\!\int\!\!\!\int_v (\rho_0\mathbf{k} + \boldsymbol{\nabla}_0\!\cdot\!\mathbf{P}) \cdot \delta\mathbf{R}dv + \int\!\!\!\int\!\!\!\int_v \mathbf{P} \cdot\!\cdot (\boldsymbol{\nabla}_0\delta\mathbf{R})^T dv =$$

$$= \int\!\!\!\int\!\!\!\int_v \mathbf{P} \cdot\!\cdot \delta\boldsymbol{\nabla}_0\mathbf{R}^T dv,$$

where (1.22) and (1.23) were used together with transposition property of $\boldsymbol{\nabla}_0$ and variation δ, namely, $\boldsymbol{\nabla}_0\delta\mathbf{R} = \delta\boldsymbol{\nabla}_0\mathbf{R}$. There is no variation of an initial state: $\delta\mathbf{r}^s = 0$. The operations $\boldsymbol{\nabla}$ and δ cannot be transposed: $\boldsymbol{\nabla}\delta\mathbf{R} = \boldsymbol{\nabla}\mathbf{r}\boldsymbol{\nabla}_0\delta\mathbf{R} = \boldsymbol{\nabla}\mathbf{r}\delta\boldsymbol{\nabla}_0\mathbf{R}$; see (1.15).

The integrand in the last integral in (1.25) is an elementary work per unit volume of initial configuration, namely, $\widetilde{\delta}a$.

Taking the motion equation (1.22) into consideration, one should substitute the *virtual* displacement $\delta\mathbf{R}$ by means of a real one $d\mathbf{R} = \mathbf{v}dt$ and write from (1.25):

$$\frac{\widetilde{\delta A}}{dt} = \int\int\int_V \rho\mathbf{k}\cdot\mathbf{v}dV + \int\int_S \mathbf{t}\cdot\mathbf{v}dS = \int\int\int_V (\rho\mathbf{k} + \boldsymbol{\nabla}\cdot\mathbf{T})\cdot\mathbf{v}dV + \int\int\int_V \mathbf{T}\cdot\cdot\boldsymbol{\nabla}\mathbf{v}^T dV =$$

$$= \int\int\int_V (\rho\frac{d\mathbf{v}}{dt}\cdot\mathbf{v})dV + \frac{1}{2}\int\int\int_V \mathbf{T}\cdot\cdot(\boldsymbol{\nabla}\mathbf{v}^T + \boldsymbol{\nabla}\mathbf{v})dV.$$

Here the symmetry of $\mathbf{T}$ was used. Evidently, the first term of last expression represents the material time derivative of kinetic energy K:

$$K = \int\int\int_V (\rho\frac{d\mathbf{v}}{dt}\cdot\mathbf{v})dV = \int\int\int_v (\rho_0\frac{d\mathbf{v}}{dt}\cdot\mathbf{v})dv = \frac{1}{2}\frac{d}{dt}\int\int\int_v (\rho_0\mathbf{v}\cdot\mathbf{v})dV$$

while the second one represents the power $\mathcal{N}$ in terms of $\mathbf{T}$ and the Piola tensor $\mathbf{P}$ respectively:

$$\mathcal{N} = \frac{1}{2}\int\int\int_V \mathbf{T}\cdot\cdot(\boldsymbol{\nabla}\mathbf{v}^T + \boldsymbol{\nabla}\mathbf{v})dV = \frac{1}{2}\int\int\int_v \mathbf{P}\cdot\cdot\boldsymbol{\nabla}\mathbf{v}^T\cdot\boldsymbol{\nabla}_0\mathbf{R}^T dv$$

Therefore the work of the external both mass and surface forces per time unit is equal to the sum of power and time changes of kinetic energy for an elastic substance.

In addition to (1.21)-(1.23) one should formulate also the energy conservation and the entropy inequality. Aiming to formulate the final equations in terms of displacement or strain components, we will discuss them below in proper cases; formal expressions of them are not of major use.

1.2   Physical and geometrical sources of nonlinearity

There are several sources of nonlinearity that affect the elastic wave propagation. Starting with a brief outlook, we mention the structural and kinematic nonlinearities. The first is well known after any physics textbook, namely, the Coulomb friction force discontinuously depends upon the body slip velocity, see also Crandall (1974) for other examples. Engelbrecht (1994) provided the examples of kinematical nonlinearity which occurred in the compound motion of solids or in rotation dynamics of rigid body.

In both cases it can be compared with the convective terms in the Navier-Stokes' equations in fluid dynamics arising from the duality of velocity with respect to the local and absolute inertial reference frames. Combined and complex nonlinearities should be mentioned, also, provided, e.g., by thermal processes influencing the mechanical stress field, or by pre-stressing of a material, or by viscous liquid- solid interface interaction.

These sources of nonlinearity will not be, however, of key interest for us in contrast to the intrinsic or material nonlinearity of elastic solids as well as the nonlinearity of deformation itself. That will be of most importance for nonlinear wave propagation theory under consideration.

Table 1.1 Stress-strain relations, after Bell, (1973)

Elasticity law	Author	Proposed usage
$\sigma = a\epsilon$	Hooke (1678)	metals
$\epsilon = a\sigma^m$	Bernoulli (1694)	general
$\sigma = a\exp(-1/\epsilon)$	Riccati (1731)	general
$\epsilon = \sigma[a + b\exp(m\sigma)]$	Poncelet (1839)	brass in tension
$\epsilon^2 = a\sigma^2 + b\sigma$	Wertheim (1847)	organic tissues
$\sigma = a\epsilon + b\epsilon^2 + c\epsilon^3 + d\epsilon^4$	Hodgkinson (1849)	cast iron
$\sigma = a\epsilon + b\epsilon^2 + c\epsilon^3$	Cox (1850)	cast iron
$\epsilon = a\sigma + b\sigma^2 + c\sigma^3$	Thompson (1891)	metals in tension
$\epsilon = \exp(a\sigma) - 1$	Hartig (1893)	India rubber

The famous Hooke's law stated the following linear stress-strain relation: $\sigma = a\epsilon$, where σ is a stress and ϵ is a strain caused by the stress and proportional to them with a coefficient a representing the intrinsic material's constant.

Bell (1973) described the detailed history of study of the nonlinearity of materials. To illustrate it we refer to his data given in Table 1.1. As it clearly seen, long before the understanding of nonlinearity of real fluid motion and the famous J. S. Russel's experiments on solitons on shallow water, the concept of nonlinearity of materials was widely discussed in previous centuries as the alternative to the Hooke law.

Therefore, for almost three centuries, elastic materials were considered as nonlinear substances, and the Hooke law seemed to be valid for very small deformations, unless otherwise stated. The nonlinear stress-strain dependence, e.g., according to the Hodgkinson law, can be caused by internal sliding of either long molecules, in polymers, or of thin layers in polycrystalline solids. Naturally it is required to analyse the microscopic structure of solids.

Thus, linear theory is well grounded on the initial stage of deformation process, when strains are small enough, see, e.g., Figure (1.5).

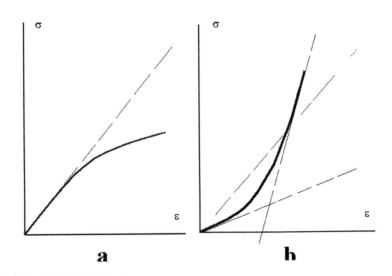

Figure 1.5: Schematic diagram for a stress-strain relation for linear (a) and nonlinear (b) elastic solids.

The postulate of elastic material existence should be added by an existence of stress 'potential' $\mathcal{V}$ as the function of the vector gradient $\boldsymbol{\nabla}_0\mathbf{R}$ or any of the deformation measures. Evidently, the potential energy of deformation will be defined as

$$U = \int\int\int_v \mathcal{V}dv$$

The variation of $\mathcal{V}$ is to be equal to the elementary work per volume unit made by external forces on the elastic material deformation. The existence of U means that an elastic body can accumulate the external forces' work when loaded. The concepts of $\mathcal{V}$ and U can be connected with the Helmholz free energy and the internal energy.

Nowadays it is known from physics of crystalline lattice that the atoms' interaction inside a solid is generally governed by the potential[2] function, that was initially introduced by Morse (1929):

$$\mathcal{V}(x) = \mathcal{V}_0[\exp(-2\alpha x) - 2\exp(-\alpha x)], \tag{1.26}$$

in the different problem of quantum mechanics of an electron motion, where both α and $(-\mathcal{V}_0)$ are constant, and the last is the potential hole depth. Later it was shown, (see, e.g., Valkering, 1978) that this highly nonlinear potential function governs precisely the nearest-neighbour interaction in an anharmonic lattice of the atoms in a diatomic molecule of a solid in a continuum limit. Exact solutions for elasticity

---

[2]A caution: a potential in atomic interaction theory differs from the potential energy by a factor, namely, by an electron (particle) charge only, and for this reason physicists use the term potential for brevity. There is no real charge in continuum mechanics, therefore one should introduce the potential energy of interaction, appropriately normalized, instead of interaction potential itself, essentially, when a model is subjected to generalize to the continuum limit.

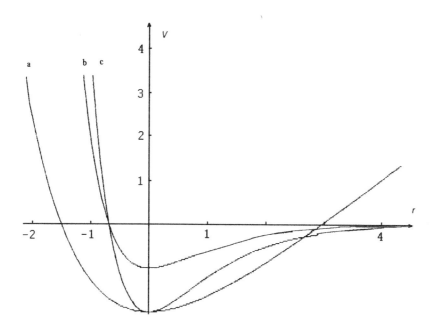

Figure 1.6: Potentials of Toda *(a)* and of Morse *(b,c)*.

problems described by this interaction potential are still unknown. However it turns out that many problems of nonlinear elasticity, introduced by different approaches, are governed by the corresponding approximations of the Morse potential.

Another lattice model is based on the potential function, prescribing an exponential interaction also:

$$\mathcal{V}(x) = (a/b)\exp(-bx) + ax, \qquad (1.27)$$

often called the Toda potential (Toda, Wadati, 1973). Evidently, it admits the linear growth of an interaction potential for $x \to \infty$, and for this reason it provides a model, which application is more restrictive from the general view point of physics of atomic lattices. Both constants are to be given as $ab > 0$, then $\mathcal{V}(x) > 0$, therefore in order to have a necessarily negative value of a potential for $x = 0$ one should add a negative constant, as shown in Fig. 1.6, that is required to be well grounded from a physical basis. Moreover for $x < 0$ the Morse potential tends to infinity much more rapidly, as it must be for the lattice model with nearest interactions taken into account.

Nevertheless, for small perturbations, a lattice will be linear with a spring constant $k = ab$, for bigger - the potential can approximated as the cubic polynomial, $\mathcal{V}(x) = abx^2/2 - ab^3x^3/3 + \dots$ with respect to the relative atomic distance.

The main advantage of the discrete lattice nonlinear model, introduced by Toda, (1981) is that it can be integrated explicitly by the inverse scattering transform, moreover in the continuum limit the corresponding equation is reducible to the famous nonlinear Boussinesq equation, having bi-directional wave solution, see, e.g., Rosenau, (1986).

It is well understood now that various hyperelastic continuum models of real materials should be based on the corresponding limits of atomic lattice models. The interaction lattice potential, and, consequently, the potential energy of deformation both will have to contain not only terms of the second order with respect to the distance between atoms but also the higher order terms. Only for sufficiently small and one-dimensional longitudinal lattice displacements , i.e., variations of interatomic distance, the energy can be approximated by means of a parabola, and strains will be linear in displacement gradients, respectively. Otherwise one should consider the often called "physical" nonlinearity of solids.

Several points can be outlined here.

Firstly, despite a common experience with erasers and child's catapults, not only for rubber-like materials but even for metals, the physical nonlinearity should be taken into account, see, e.g., Parker, (1984). For copper and iron, brass and lead, the next order elasticity values are much larger then the Lamé constants, and therefore may provide the relative increase in potential energy up to 20% for very small values of the second order terms in tensor $\mathbf{C}$, as shown by Pleus and Sayir, (1983). Similar behaviour of granular and porous materials was discussed also by Wright (1984) and Nesterenko (1992).

Finally one can conclude that the expression for potential energy density should remain fully nonlinear in the statement of dynamics elasticity problem, unless a correctness of any particular linear approximation can be proven. In other words, the condition for small displacement gradient $\mid \partial U_i / \partial x_j \mid \, << 1$ does not necessarily lead to the relatively small addition in strain quantities, $(u_{nonlin} - u_{lin})/u_{lin} << 1$, see (Parker, 1984), that was typical for linear theory of elasticity, and the reason for discrepancy consists, at least, of possible large intervals of time.

For usage of variational principle it is assumed to be given the volume density $\mathcal{U}$ of internal energy as the function of entropy $\mathcal{S}$ and invariants I_k of any tensor of deformation, all are defined for an adiabatic process of deformation. Another thermodynamic potential can be considered also, e.g., the Helmholz free energy $\mathcal{F}$ is useful for isothermic (thermoelastic) deformation problems consideration, see (Engelbrecht, 1983; Maugin, 1994).

1.2.1 Elastic potentials and moduli

It is a complicated task to describe concisely main ideas of mathematical and physical treatment of nonlinear elasticity problems. The best solution seems to refer from the very beginning to well known books by Antman (1995); Lurie, (1980); Maugin, (1988); Eringen and Suhubi, (1975); Engelbrecht, (1983), and some others, inspiring any diligent reader with a not so simple introduction and recent developments in the topic. Nevertheless, concentrating on nonlinear guided wave problems, we will have to repeat general viewpoints of nonlinear elasticity theory[3]. The introductory notes given here are necessary to simplify an explanation and will be used to reveal the object under study, and to refer to further reading, and to discuss some important problems besides an explanation.

---

[3]Brief repetition never can be a substitute for the genuine treatise on the theory, and we may recommend for newcomers to start with the books in nonlinear elasticity.

Our aim is to introduce and to apply the Hamilton principle in order to derive the (nonlinear) equation governing the wave propagation in a nonlinearly elastic solid wave guide from the first principles. To do it one has to introduce the action functional defined on the vector **R** in actual configuration.

We will assume that a nonlinearly elastic material "does not know its future", and has no viscous and/or viscoelastic features, that provide to nominate deformation process as the reversible one.

Writing again the finite deformation tensor from (1.18),

$$\mathbf{C} = [\boldsymbol{\nabla}\mathbf{U} + (\boldsymbol{\nabla}\mathbf{U})^T + \boldsymbol{\nabla}\mathbf{U}\cdot(\boldsymbol{\nabla}\mathbf{U})^T]/2,$$

in terms of tensors $\boldsymbol{\nabla}\mathbf{U}$ and $(\boldsymbol{\nabla}\mathbf{U})^T$, i.e., the vector-gradient and its transpose, respectively, of the displacement vector $\mathbf{U} = \{U_k\}$, we introduce into consideration possibly large geometrical variations of initial configuration of an elastic solid. In other words, the third term in **C** reflects small but finite, however *not infinitesimal deformations* (as it is in linear elasticity), and strains, resulted from them, respectively.

When the deformation process is adiabatic, the internal energy U can be expressed as a partial sum of power series with respect to $I_k(\mathbf{C}^n)$, $k = 1, 2, 3$, introduced in (1.19),

$$I_1(\mathbf{C}) = tr\mathbf{C}, \quad I_2(\mathbf{C}) = [(tr\mathbf{C})^2 - tr\mathbf{C}^2]/2, \quad I_3(\mathbf{C}) = \det \mathbf{C}, \tag{1.28}$$

and with coefficients, depending on the elastic moduli (λ, μ) of second order (the classic Lamé constants), and those of third order ν_k,

$$\rho_0 U = (\lambda/2)I_1^2(\mathbf{C}) + \mu I_1(\mathbf{C}^2) + (\nu_1/6)I_1^3(\mathbf{C}) + \nu_2 I_1(\mathbf{C})I_1(\mathbf{C}^2) + (4\nu_3/3)I_1(\mathbf{C}^3) + \dots \tag{1.29}$$

and so on. Using the standard technique, see (Lurie, 1980), to recalculate the invariants $I_k(\mathbf{C}^n)$ of power of **C** in terms of $I_k(\mathbf{C})$, e.g.,

$$I_1(\mathbf{C}^2) = I_1^2(\mathbf{C}) - 2I_2(\mathbf{C}); \quad I_1(\mathbf{C}^3) = 3I_3(\mathbf{C}) + I_1^3(\mathbf{C}) - 3I_1(\mathbf{C})I_2(\mathbf{C}),$$

this equation can be expressed in terms of $I_k(\mathbf{C})$ only.

Another power series expansion was proposed by Landau and Rumer (1937), see also (Landau, Livshitz, 1987). They formally introduced two square (u_{ik}^2, u_{ll}^2) and three cubic $(u_{ll}^3, u_{ll}u_{ik}^2, u_{ik}u_{il}u_{kl})$ scalars defined by components of a symmetric second rank tensor **u**. Then the most general scalar containing square and cubic terms with respect to u_{ik} will have the form

$$\mathcal{E} = \mu u_{ik}^2 + (K/2 - \mu/3)u_{ll}^2 + (A/3)u_{ik}u_{il}u_{kl} + B\,u_{ll}u_{ik}^2 + (C/3)u_{ll}^3 \tag{1.30}$$

where the bulk modulus K and the higher order moduli (A, B, C) were introduced. Substituting here (1.18) will result in the elastic energy expression with terms up to the third order in $(\partial u_i/\partial x_k)$.

Terms in **C** proportional to $\boldsymbol{\nabla}\mathbf{U}\cdot(\boldsymbol{\nabla}\mathbf{U})^T$ are usually referred to as the *geometrical nonlinearity*, whereas all terms in Π except the first two are named as the *physical nonlinearity*. We should note that the subdivision is a colloquial one, because of *intrinsic* *overlap* between the two, in part, due to the values of invariants I_k . Moreover, in general, we should emphasize that both geometrical and physical nonlinearities must be taken simultaneously into consideration, i.e., square terms with respect to $\nabla \mathbf{U}$ in $\mathbf{C}$, cubic and higher order terms with respect to strains $\partial u_i / \partial x_k$ in components of Π, even for metals under finite, but not infinitesimal deformation, to say nothing of rubber, polymers, and composites. We can show the mutual influence, following Lurie, (1980), by writing down the stress tensor $\mathbf{T}$ in the initial configuration in the form

$$\mathbf{T} = (1 - \theta)\mathbf{T}^0 + 2\mathbf{e}\mathbf{T}^0 + \mathbf{T}^1,$$

where $\mathbf{T}^0 = \lambda\theta\mathbf{E} + 2\mu\mathbf{e}$ is the stress tensor for linear elasticity, $\mathbf{E}$ is the unit tensor, $\mathbf{e}^0 = [\nabla\mathbf{U} + (\nabla\mathbf{U})^T]/2$ is the linear deformation tensor, whereas $\mathbf{T}^1 = a\mathbf{E} + b\mathbf{e}^0 + \mu(\nabla\mathbf{U})^T\nabla\mathbf{U} + n(\mathbf{e}^0)^2$ is the additional tensor and coefficients $a = a(\lambda, \mu; l, m, n)$ and $b = b(\lambda, \mu; l, m, n)$ depend on elasticity. It is evident now that $\mathbf{T} \neq \mathbf{T}^0$, even for $l = m = n = 0$, i.e., without any "physical" nonlinearity. Conversely, if, following H. Kauderer (1958), one could use a partial sum for Π of any order with invariants defined for the tensor $\mathbf{e}^0$ instead of $\mathbf{C}$, i.e., $I_k(\mathbf{C}) \neq I_k(\mathbf{e}^0)$, that could result in the loss of some terms of second order with respect to $\nabla\mathbf{U}$, e.g., of $I_1(\mathbf{e}^0)T^0(\mathbf{e}^0)$ and so on; see Lurie (1980) for details.

The next step in construction of the Lagrangian density consists in a proper choice of approximation of Π through a partial sum of a series with respect to I_k, and with coefficients, depending on the elastic moduli (λ, μ) of second order, and of third order (ν_i), and of 4th order (γ_i), and so on. We describe briefly the most important models for nonlinearly elastic (or hyperelastic, as they are often called) materials and write for the beginning the Murnaghan energy expansion up to 9-constant approximation:

$$
\begin{aligned}
\Pi \;=\; & (\lambda + 2\mu)I_1^2/2 - 2\mu I_2 + (l + 2m)I_1^3/3 - 2m I_1 I_2 + n I_3 + \\
& + \gamma_1 I_1^4 + \gamma_2 I_2^2 + \gamma_3 I_1 I_3 + \gamma_4 I_2^2 + ...
\end{aligned}
\tag{1.31}
$$

Here the following Murnaghan's moduli were introduced: $l = \nu_1/2 + \nu_2$, $m = \nu_2 + 2\nu_3$, $n = 4\nu_3$, see (1.29).

All coefficients of one approximation, e.g., from (1.30), can be transformed into those of another one and vice versa. Note also, that any strain should be sufficiently small for Π to be a convex function with respect to the corresponding longitudinal strain. Otherwise the approximation of Π by a partial sum is not valid.

The Murnaghan approximation (1.31) for Π seems to be the most useful for compressible nonlinearly elastic materials under small deformations, but not the unique one.

For description of any rubber-like (compressible) material, the Blatz-Ko model (Blatz, 1969) is quite accurate, and Π has the form:

$$\Pi = \mu\kappa[i_1 - 3 + (i_3^{-a} - 1)]/2 + \mu(1 - \kappa)[i_2/i_3 - 3 + (i_3^a - 1)]/2, \tag{1.32}$$

where $a \equiv \nu/(1 - 2\nu)$, and κ is the only modulus of the third order ($0.5 < \kappa < 1$), whilst all the invariants $i_k = i_k(\mathbf{G})$ were defined here with respect to the Green deformation measure tensor $\mathbf{G} = 2\mathbf{C} + \mathbf{E}$, and can be recalculated as $i_1 = 2I_1 + 3, i_2 = 4I_1 + 4I_2 + 3, i_3 = 2I_1 + 4I_2 + 8I_3 + 1$, where $I_k = I_k(\mathbf{C})$.

As for incompressible or near-incompressible materials (e.g., PVC and rubber), one should take into account the restriction $\det\{\mathbf{G}\} = 1$, which results in a simple formula for Mooney's material (Mooney, 1940) :

$$\Pi = E[(3 - \kappa)I_1 + 2(1 - \kappa)I_2]/6, \ |\kappa| < 1, \tag{1.33}$$

where E is the Young's modulus, and the relation $\mu = E/3$ is now valid. Recently the Mooney-Rivlin model was used also (Cohen and Dai, 1993) to study long nonlinear axisymmetric waves in compressible rods.

For Treloar's model (Treloar, 1958), the simplest one, we have $\mu = E/3$ and in addition $\nu = 1/2$, and as a result

$$\Pi = EI_1/3, \tag{1.34}$$

note that $I_k = I_k(\mathbf{C})$ in (1.33),(1.34), i.e., the nonlinearity is due to geometry of deformation only.

The simplicity of these relations does not result, unfortunately, in the only single equation for nonlinear waves; from the general view point, the nonlinear differential equations remain coupled.

Wave problems with higher order nonlinearities proportional to u^3 or u^5 were studied recently in (Clarkson, LeVeque and Saxton, 1986); here we underline only that it requires to take into consideration also the following nonlinear term $\epsilon(cu^2 + cu^2)$ and corresponding dispersive terms, when a highly nonlinear elasticity problem is studied.

1.3 Compressibility, dispersion and dissipation in wave guides

The low order dispersion term sufficiency for the soliton solution existence was demonstrated by Rybak, Skrynnikov (1990), which dealt with a problem on soliton propagation in a uniformly curved elastic rod. It resulted in the following 1+1D equation:

$$u_{tt} - c^2 u_{xx} = \frac{\beta}{2\rho}(u^2)_{xx} - \frac{c^2}{R}u$$

where β is a nonlinearity coefficient, ρ is density, R is the curvature radius of the middle line of the rod and c^2 is the linear wave velocity. They found the solitary wave solution almost similar to the standard one.

Another interesting example of exact solution was found to the 3D solitary wave propagating in a system devoid of potential energy, namely, to the solitary wave in inextensible in elastic helix (Slepyan et al. 1995). Solitary waves found in this system were governed by a 2D vector equation in complex variable plane as follows:

$$(1 - f)\mathbf{r}'' - f'\mathbf{r}' - 2i\lambda\mathbf{r}' - \lambda^2\mathbf{r} = 0$$

where **r** is the vector radius in a cross section of the helix, f is proportional to the constant tension force applied to the helix, γ is the initial angle between a fiber and a axis of the helix, and primes denote the differentation in coordinate along the fiber. This equation contains the low order dispersion and the dissipative term proportional to the first derivate of the unknown. The general solution to it may be found using an approach described in Chapter 3.

In presence of gravity there are no totally free wave guides, which should be supported, at least, mechanically, in one end as a console or even embedded into a surrounded external medium. In space one could imagine a free wave guide, however any disturbance would provide its motion as a whole, not a wave, that is of minor interest. Due to contact an external medium will cause an energy flux *out* of a wave guide (i.e., dissipation from the view point of a guided wave) or an energy influx *into* a wave guide from an (active) external medium. Typical sources of dissipation are kinematic viscosity of a wave guide material and possible energy dissipation due to interaction along a contact lateral surface. Active external media are well known in biology, chemistry, seismic problems, etc.

As a rule, a wave problem considered in solid wave guide includes a stress-free boundary condition at the lateral surface. However the longitudinal wave propagation leads to the non-zero transversal displacement wave, hence to the deformation of a lateral surface of the wave guide being restricted by a contact with an external medium. When this medium is air with vanishing contact resistance, the only problem is how to take into account a force in movable or fixed support, while for an embedded wave guide one should consider a model of the surface interaction. This interaction results in an energy flux to or from a wave guide.

Nonlinear wave problems with energy influx are typical for many applications. Some of them were analysed, e.g., by Engelbrecht and Peipman (1992) in seismic layer problems and by Engelbrecht (1991) in problems of soft biological tissues. The final form of corresponding equation may be similar to the KdV-Burgers equation or the FitzHugh-Nagumo equation or the nonlinear reaction-diffusion equation; details of derivation can be found in (Scott, 1999; Samsonov, Gursky, 1999). In Chapter 3 we will discuss a new approach to find some exact solutions to these problems.

A rod embedded in a prestressed external medium, a blood vessel, a bone or a nerve fiber compressed by a muscle, construction in a chemically active, or pre-heated, or shocked environment seem to be proper examples of such contact problems under consideration. Conversely, an wave energy can dissipate through a lateral surface into an external medium, say, during hammering of a nail. Finally a medium can be passive: an interaction does not provide any change of energy of internal bulk wave in a wave guide.

Here we consider a pure mechanical wave motion originated from nonlinear elasticity in which the energy exchange is to be taken into account, and for this reason we will start with a brief description of some models of interaction of a wave guide and an external medium, and firstly, with some elastic and viscoelastic foundation models.

The simplest representation of a continuous elastic foundation (and for our purposes, of interaction between a surface of a wave guide and a medium) has been provided by Winkler (1867). The interaction was described as the only transversal motion, e.g., of compressed springs:

$$F(\mathbf{x}, t) = -k_1 u_k \qquad (1.35)$$

where F is the pressure and u_k are deflections of foundation surface. Therefore the longitudinal motions of a wave guide surface do not interact with them, and it is not of considerable interest for our dynamical problems; such a foundation is equivalent, in a sense, to a liquid foundation.

There is a large class of materials that, say, behaviour cannot be described by a Winkler-type interaction. According to the improved model, proposed by Pasternak (1954), the shear interaction between the spring elements in motion should be taken into account that leads to:

$$F(\mathbf{x}, t) = -k_1 u_2 + k_2 \frac{\partial^2 u_2}{\partial x_1^2}. \qquad (1.36)$$

where x is the longitudinal coordinate along the axis, and k_1, k_2 are the stiffness coefficients of the medium with respect to compression and to shear, respectively. Similarly, in order to add an interaction between spring elements, Filonenko-Borodich (1940) proposed to connect the top ends of springs with an expanded membrane. The final expression looks like (1.36) where k_2 becomes the constant of tension field.

Later Reissner (1958) begun with the continuum equations and suggested that in-plane stresses in the foundation are zero and obtained an expression reducible to written above when a distributed lateral load on the surface is assumed to be linear with a distance along the axis. However, the zero shear stresses in plane mean that other shears are independent of a variable normal to the interaction surface, and therefore, are constant throughout the depth of a foundation. That seems not to be valid for thick layers of external medium.

One can conclude that the model of elastic interaction based on a continuum contact problem on a surface is not yet well developed, and for our 'leading order' consideration, the usage of Pasternak model seems to be reasonable.

When the energy exchange between the rod and the medium is considered, i.e, there is either an energy influx from an active external medium into the wave or a dissipation of a deformation wave in the viscous external medium, the extension of the Pasternak model is necessary. This is achieved by adding linear viscous elements to elastic ones either in parallel or in series. Such problems were considered by Kerr (1964). He found that the force $F(\mathbf{x}, t)$ can be expressed as it follows from his model:

$$F(\mathbf{x}, t) = -k_1 u_2 - \eta.\frac{\partial u_2}{\partial t} \qquad (1.37)$$

where η is either viscocompressibility or the energy influx coefficient, positive for the viscous external medium and negative for the active one. From the physical viewpoint, it means that the viscocompressibility related to shear deformations of the elements of foundation was taken into account. This model will be considered in detail below in Section 2.5 (see also Fig.2.2) as the proper one for nonlinear waves in a rod embedded in an elastic or viscoelastic medium.

Chapter 2

A mathematical description of the general deformation wave problem

A mathematical description of the general deformation wave problem is based on both nonlinear elasticity theory and nonlinear wave theory in condensed matter, in particular, the theory of solitons. Both topics are hard to be observed even briefly; the last is growing rapidly during the last several decades, and one has no chance to mention even the most important contributions. We have to limit references here to those which material will be refered to directly, e.g., books and reviews in nonlinear elasticity by Murnaghan (1954), Treloar (1958), Bland (1969), Lurie (1980), Shield (1983), in nonlinear elastic waves theory by Engelbrecht (1983, 1991), Maugin (1994), as well as those written by Jeffrey and Kawahara (1982), Nayfeh (1973), Cole (1968), Lomov (1981) and by Kodama and Ablowitz (1981), to note only several useful approaches to perturbation methods aimed at the applications in nonlinear mathematical physics and engineering. There are many unsolved nonlinear guided wave problems, and our reference list is in no way complete. However, main ideas were mostly settled in the last decade, and the nonlinear elastic wave theory may be formalised now.

2.1 Action functional and the Lagrange formalism

Define the action functional S in the form

$$S = \int_{t_0}^{t_1} \left(\int_{\Omega} \mathcal{L}(U, U_t, U_x, \ldots x, t) d\Omega + \int_{\partial\Omega} \mathbf{F}_k \mathbf{U}_k d(\partial\Omega) \right) |_{t=t_0} dt, \qquad (2.1)$$

where $\mathcal{L}(U, U_t, U_x, \ldots x, t)$ is the Lagrangian density per unit volume Ω, t is time, $\mathbf{U} = \{U, V, W\}$ is the displacement vector, defined in the Lagrangian co-ordinates $\{x_k\}$, $k = 1, 2, 3$. The second term in (2.1) expresses the work of the external forces $\mathbf{F}_k$, which occurs on the (lateral) surface $\partial\Omega$ of the wave guide as a consequence of the reaction of the external medium, and results from bounded transverse displacement. The initial configuration (for $t = t_0$) of an elastic volume is assumed to be the natural one, thermal features are not considered.

Using the Hamilton principle,

$$\delta\mathcal{L} + \widetilde{\delta}A = 0 \tag{2.2}$$

one can obtain a set of basic Euler equations, i.e., the coupled partial differential equations governing the wave propagation problem. Indeed, the first variation leads to the Euler-Lagrange equations, which in a simple case of the one displacement U can be written in general form as:

$$\frac{\partial\mathcal{L}}{\partial U} - \left[\frac{\partial}{\partial t}\frac{\partial\mathcal{L}}{\partial U_t} + \frac{\partial}{\partial x}\frac{\partial\mathcal{L}}{\partial U_x}\right] + \frac{\partial^2}{\partial x^2}\frac{\partial\mathcal{L}}{\partial U_{xx}} + \frac{\partial^2}{\partial x\partial t}\frac{\partial\mathcal{L}}{\partial U_{xt}} + \cdots - \frac{\partial^3}{\partial x^3}\frac{\partial\mathcal{L}}{\partial U_{xxx}} - \cdots = 0 \tag{2.3}$$

Again, δA is in no way 'the variation of work', that cannot be defined, but only an *elementary work* produced by external forces with respect to virtual motion considered. For example, one should consider possible variations of elasticity and geometry of a wave guide in space and time that will give an awkward expression for energy even before variations, and for this reason analytical consideration has to be based on sophisticated treatment of the problem. Guided waves in a hyperelastic material may be short or long, providing either small finite or large deformations in dependence of (in)compressibility of the material, not to say about uni- or bidirectional propagation in one- or two-dimensional wave guide, etc., etc. The invariance features of the tensor formulation of dynamic problems is very useful in general analysis but rather restrictive in application to particular nonlinear wave propagation problem or real experiments, where one has to reduce the statement to the p.d.e. written with respect to strain components-and measure them.

Main assumptions will be about geometry of a wave guide-and we shall consider rods in cylindrical co-ordinates and thin plates in cartesian co-ordinates, and of elasticity of it, that will lead to simplifications based on either compressibility and small deformations or on incompressibility of a material.

When the deformation process is assumed to be adiabatic, the function $\mathcal{L}$ in the material (Lagrangian) variables can be determined as the difference of the kinetic energy density T and the volume density Π of potential energy, as follows:

$$\mathcal{L} = T - \Pi = \rho(\partial U/\partial t)^2/2 - \rho\Pi(I_k), \tag{2.4}$$

where ρ is the material density, and $I_k = I_k(\mathbf{C})$ are the invariants of the Cauchy-Green finite deformation tensor $\mathbf{C}$:

$$\mathbf{C} = [\nabla\mathbf{U} + (\nabla\mathbf{U})^T + \nabla\mathbf{U}\cdot(\nabla\mathbf{U})^T]/2, \tag{2.5}$$

where the upper index T denotes the diad transpose: $\mathbf{ab}^T \equiv (\mathbf{ab})^T \equiv \mathbf{ba}$. These invariants may be written as

$$I_1(\mathbf{C}) = \text{tr }\mathbf{C}, \ \ I_2(\mathbf{C}) = [(\text{tr }\mathbf{C})^2 - \text{tr }\mathbf{C}^2]/2, \ \ I_3(\mathbf{C}) = \det\mathbf{C}. \tag{2.6}$$

Note that $\nabla\mathbf{U}$ and $(\nabla\mathbf{U})^T$ are the vector-gradient (i.e., the tensor) and its transpose, respectively, of the displacement vector $\mathbf{U} = \{U, V, W\}$. We use tensors, defined in the reference configuration of an elastic system (i.e., for $t = t_0$) , and omit the subscript 0 for brevity.

We are interested in the simplest (but not trivial) statement of the nonlinear problem, involving in the 1+1D case a single p.d.e. with respect to the one unknown function, as the natural generalization of the standard hyperbolic equation, used for the wave propagation description in a linearly elastic wave guide. Generalisation for the 2D wave guides will be considered also, and our main goal will be in the development of the theory instructive for the solitary wave generation and observation in elastic wave guides.

Firstly the nonlinear waves in a hyperelastic compressible infinite rod will be considered. Introducing the unit vectors in a cylindrical co-ordinate system as $(\mathbf{e}_x, \mathbf{e}_r, \mathbf{e}_\phi)$ and a displacement vector $\mathbf{U} = U\mathbf{e}_x + V\mathbf{e}_r + W\mathbf{e}_\phi$ together with a vector-gradient $\boldsymbol{\nabla} = \partial_x\mathbf{e}_x + \partial_r\mathbf{e}_r + (1/r)\partial_\phi\mathbf{e}_\phi$, one can write the finite deformation tensor $\mathbf{C}$ in the form (2.5) and, calculating

$$\boldsymbol{\nabla}\mathbf{e}_x = 0; \boldsymbol{\nabla}\mathbf{e}_\phi = -\frac{1}{r}\mathbf{e}_\phi\mathbf{e}_r; \boldsymbol{\nabla}\mathbf{e}_r = \frac{1}{r}\mathbf{e}_\phi\mathbf{e}_\phi, \tag{2.7}$$

obtain the diad

$$\boldsymbol{\nabla}\mathbf{U} = U_x\mathbf{e}_x\mathbf{e}_x + U_r\mathbf{e}_r\mathbf{e}_x + V_x\mathbf{e}_x\mathbf{e}_r + V_r\mathbf{e}_r\mathbf{e}_r + \frac{V}{r}\mathbf{e}_\phi\mathbf{e}_\phi. \tag{2.8}$$

Here $U\boldsymbol{\nabla}\mathbf{e}_x = 0$ was taken into account, and the *first assumption* was introduced, namely, torsion is assumed to be negligible: $W = 0$, during any of deformations of a wave guide. Without torsion, the components of the displacement vector $\mathbf{U}$ are independent of φ and equal to

$$\mathbf{U} = (U, V, 0); \ U = U(x, r, t), \ V = V(x, r, t),$$

The Lagrangian (2.4) is transformed now to the form:

$$\mathcal{L} = T - \Pi = \frac{\rho}{2}\left[U_t^2 + V_t^2\right] - \rho\Pi[I_k(\mathbf{C})], \tag{2.9}$$

Similarly, we calculate the tensor

$$\boldsymbol{\nabla}\mathbf{U} \cdot \boldsymbol{\nabla}\mathbf{U}^T = U_x^2\mathbf{e}_x\mathbf{e}_x + 2U_rU_x\mathbf{e}_r\mathbf{e}_x + V_x^2\mathbf{e}_x\mathbf{e}_x + U_r^2\mathbf{e}_r\mathbf{e}_r + 2V_xV_r\mathbf{e}_x\mathbf{e}_r + V_r^2\mathbf{e}_r\mathbf{e}_r + (\frac{V}{r})^2\mathbf{e}_\phi\mathbf{e}_\phi \tag{2.10}$$

Therefore for the tensor components it yields: $C_{x\phi} = C_{r\phi} = 0$, and the tensor $\mathbf{C}$ can be written now in the form:

$$\mathbf{C} = \begin{vmatrix} C_{xx} & C_{xr} & 0 \\ C_{rx} & C_{rr} & 0 \\ 0 & 0 & C_{\phi\phi} \end{vmatrix} \tag{2.11}$$

while for its square one has:

$$\mathbf{C}^2 = \begin{vmatrix} C_{xx}^2 + C_{rx}^2 & \dots & \dots \\ \dots & C_{xr}^2 + C_{rr}^2 & \dots \\ \dots & \dots & C_{\phi\phi}^2 \end{vmatrix} \tag{2.12}$$

where dots indicate components, which will not be used further. After calculation of $\text{tr } \mathbf{C}^2 = C_{xx}^2 + 2C_{rx}^2 + C_{rr}^2 + C_{\phi\phi}^2$, the invariants in (2.6) become in terms of components:

$$I_1(\mathbf{C}) = C_{xx} + C_{rr} + C_{\phi\phi} \tag{2.13}$$

$$I_2(\mathbf{C}) = C_{xx}C_{rr} + C_{\phi\phi}C_{xx} + C_{rr}C_{\phi\phi} - C_{rx}^2 \tag{2.14}$$

$$I_3(\mathbf{C}) = C_{\phi\phi}(C_{xx}C_{rr} - C_{rx}^2) \tag{2.15}$$

where the components of $\mathbf{C}$ are written now as:

$$C_{xx} = U_x(x,r,t) + (1/2)[U_x^2(x,r,t) + V_x^2(x,r,t)],$$

$$C_{rr} = V_r(x,r,t) + (1/2)[U_r^2(x,r,t) + V_r^2(x,r,t)],$$

$$C_{\varphi\varphi} = V/r + (1/2)(V/r)^2,$$

$$C_{rx} = (1/2)[U_r(x,r,t) + V_x(x,r,t) + U_r(x,r,t)U_x(x,r,t) + V_r(x,r,t)V_x(x,r,t)],$$

and subscripts r and x with displacement components denote differentiations. Invariants of $\mathbf{C}$ follow from (2.6) in the explicit form:

$$I_1 = U_x + V/r + V_r + (1/2)[U_x^2 + V_x^2 + V^2/r^2 + U_r^2 + V_r^2], \tag{2.16}$$

$$
\begin{aligned}
I_2 = {} & [V/r + V^2/(2r^2)][U_x + (U_x^2 + V_x^2)/2 + V_r + (U_r^2 + V_r^2)/2] - \\
& (1/4)(U_r + V_x + U_r U_x + V_r V_x)^2 \\
& + [U_x + (U_x^2 + V_x^2)/2][V_r + (U_r^2 + V_r^2)/2] + \ldots,
\end{aligned} \tag{2.17}
$$

$$
\begin{aligned}
I_3 = {} & [V/r + V^2/(2r^2)]\{-(1/4)(U_r + V_x + U_r U_x + V_r V_x)^2 + [V_r + \\
& (U_r^2 + V_r^2)/2][U_x + (U_x^2 + V_x^2)/2]\} \ldots.
\end{aligned} \tag{2.18}
$$

We omitted here for brevity most of the higher order terms; the full expressions for invariants in components are rather cumbersome and may be found in Appendix.

2.2 Coupled equations of long wave propagation

Using the expressions for invariants (2.16), (2.17), (2.18) for direct calculation of the Murnaghan model expression for potential energy in absence of torsion and substituting them into the Lagrangian density (2.9), we obtain as the Euler-Lagrange equations the *coupled* highly nonlinear equations for waves of displacements U and V, propagating in a rod, made of the Murnaghan material, and without torsion:

$$
\begin{aligned}
\rho U_{tt} - (\lambda + 2\mu)U_{xx} = {} & \lambda V_x + \mu U_{rr} + (\lambda + \mu)V_{xr} + 6(\lambda + 2\mu)U_x U_{xx} \\
& + (\lambda + 2\mu + m)\left[(U_r U_x)_r + (U_r V_r)_r + V_x V_{xx}\right] \\
& + (4\lambda + 7\mu - 3m)V_r V_{rx} + (m + \mu)V_{rr}V_x \\
& + \frac{3\lambda + 6\mu - 6m + n}{r}(V_r V)_x + (m - \frac{n}{2})(\frac{VV_x}{r})_r \\
& + 2(2\lambda + 3\mu - 2m)(U_x V_r)_x + (m + \mu)\left[(U_x V_x)_r + (U_r V_x)_x\right] \\
& + \frac{2\lambda + 2m - n}{2}(\frac{VU_r}{r})_r + \frac{4\lambda + 6\mu - 4m}{r}\left[(VU_x)_x + \frac{VV_x}{r}\right]
\end{aligned}
$$

$$(2.19)$$

$$
\begin{aligned}
\rho V_{tt} - (\lambda + 2\mu)V_{rr} = {} & \lambda\frac{U_x}{2} + (\lambda + \mu)U_{xr} + \frac{2m - n + 2\lambda}{4r}V + \mu V_{xx} + 2\mu\frac{V}{r^2} + 2\lambda\frac{V_r}{r} \\
& + (\lambda + 3\mu - 2m)\left[\frac{V^2}{r^3} + \frac{V_r^2}{r}\right] + (4\lambda + 7\mu - 3m)U_x U_{xr} \\
& + \frac{2\lambda + 3\mu - 2m}{r}\left[U_x^2 + 2r(V_r U_x)_r\right] \\
& + (\lambda + 2\mu + m)\left[U_r U_{rr} + 2V_x V_{xr} + V_r V_{xx} + (U_x V_x)_x\right] \\
& + \frac{4\lambda + 6\mu - 4m}{r^2}\left[6VV_r + VU_x + r^2(\frac{VV_r}{r})_r\right] \\
& + (3\lambda + 6\mu - 6m + n)\left[\frac{V_r U_x}{r} + (\frac{VU_x}{r})_r\right] \\
& + \frac{2m - n}{2r}(2U_r V_x + VU_{rx}) + \frac{2\lambda + 2m - n}{4r}\left[U_r^2 + 2(VV_x)_x\right] \\
& + (m + \mu)\left[(V_r U_r)_x + (U_r V_x)_r + U_{xx}U_r\right]
\end{aligned}
$$

$$(2.20)$$

Characteristics of material were assumed to be constant here for simplicity. However, even in this case, the equations look awkward and may be used mostly for further simplifications and formal analysis and for direct numerical simulations.

For the 9-constant Murnaghan elasticity theory the governing equations for displacements or strains produced by nonlinear elastic waves in a rod are even more cumbersome.

To reduce the problem to the one-dimensional one we shall simplify all these equations using either the relationships between displacement components (U, V), or introducing the small strain limit, or both.

2.3   One-dimensional quasi-hyperbolic equation

Plan of further simplifications to be done is the following: we first introduce the equation derived under maximal assumptions, which is based mostly on the physics of waves, afterwards we shall confirm these assumptions using asymptotic expansions and eventually generalise the one-dimensional nonlinear equation to the problem with variable coefficients, and to the dissipative wave propagation problems.

Evidently, to extract main features of nonlinear wave propagation in a wave guide one should simplify the problem as much as possible; generalizations may be the next step following some primary non-trivial results. It was A. E. H. Love (1927), who had taken considerable attention to the influence of the finite wave guide cross section on wave propagation in *linearly* elastic rod. Following the analysis made by Pochhammer (1876) and Chree (1889) half a century before him, he proposed the *assumptions* for the components U and V described in terms of a new function U_1, independent of r,

$$U \equiv U_1(x,t), \ V = -r\nu\frac{\partial U_1(x,t)}{\partial x} \tag{2.21}$$

which express, respectively,

i) the planar cross section hypothesis,

that is, a cross section remains plane and perpendicular to the rod axis after deformation,

and

ii) the Poisson effect,

that is well known from physics: a conventionally elastic rod is thinned down by tension, while the Poisson ratio ν is a measure of thinness.

Substitution of (2.21) into (2.16- 2.18) reduces the expressions for invariants to the following forms:

$$I_1 = (1 - 2\nu)U_{1,x} + (1/2)(1 + 2\nu^2)[U_{1,x}^2 + \nu^2 r^2 U_{1,xx}^2]; \tag{2.22}$$

$$I_2 = \nu(\nu - 2)U_{1,x}^2 - \nu(1 - \nu + \nu^2)U_{1,x}^3 - (1/4)\nu^2 r^2 U_{1,xx}^2 + O(U_{1,x}^4), \tag{2.23}$$

$$I_3 = \nu^3 U_{1,x}^3 + O(U_{1,x}^4). \tag{2.24}$$

the last two expressions are truncated in $O(U_{1,x}^4)$, that is sufficient for the description of nonlinear wave of small strain in the leading order of approximation. Having these relationships one can calculate the potential energy density for elastic potentials described in Subsection 1.2.2. Naturally, the variation of (2.4) and tentative limit of small $U_{1,x}$ are not to be transposed, then variation should be done first.

In addition, it is assumed that:

iii) long waves only are considered, with typical length $\Lambda >> R$;

iv) linear strain components are finite (not infinitesimal) with magnitudes A sufficiently small: $A << 1$.

The whole set of these assumptions is a bit restrictive, however all four of them were allowed to formulate in the eighties the simplest and rich in content nonlinear wave propagation problem for different hyperelastic solids. Some generalizations in the statement will be given below.

2.3.1   Derivation of the Doubly Dispersive Equation (DDE)

We show first in detail the direct and explicit derivation of the DDE governing the wave propagation in the nonlinearly elastic inhomogeneous Murnaghan's rod with the circular non-uniform cross section under the assumptions *i)-iv)*, i.e., in the simplest non-trivial case.

In rod's theory, the relationship $V = -\nu r U_x$, where $U = U(x,t)$, and the plane cross section hypothesis (2.21) lead to simple expressions for invariants, see (A1.5)-(A1.7) in Appendix. Therefore, the Lagrangian density $\mathcal{L}$ can be written explicitly as:

$$\mathcal{L} = \frac{\rho S}{2} U_t^2 + \frac{\rho}{4\pi} \nu^2 S^2 U_{tx}^2 - \frac{ES}{2} \left(U_x^2 + \frac{\beta}{3E} U_x^3 + \frac{\nu^2 S}{4\pi(1+\nu)} U_{xx}^2 \right), \qquad (2.25)$$

where the cross section area is $S = \pi R^2(x,t)$, R is the radius of it, and the only nonlinearity coefficient is introduced now as $\beta = 3E + 2l(1-2\nu)^3 + 4m(1+\nu)^2(1-2\nu) + 6n\nu^2$. The Euler equation (2.3) for the displacement U in the non-uniform inhomogemeous rod yields:

$$(\rho S U_t)_t - (ESU_x)_x = \frac{\partial}{\partial x} \left[\frac{S\beta}{2} U_x^2 + \frac{\rho \nu^2 S^2}{2\pi} U_{ttx} - \frac{\partial}{\partial x} \left(\frac{\nu^2 ES^2}{4\pi(1+\nu)} U_{xx} \right) \right], \qquad (2.26)$$

and it constitutes the inhomogeneous DDE in terms of displacement.

The next step of reduction is to assume that ρ and S are independent of time t. It results in the equation:

$$U_{tt} - \frac{1}{\rho S} (ESU_x)_x = \frac{1}{2\rho S} \frac{\partial}{\partial x} \left[S\beta U_x^2 + \frac{\rho \nu^2 S^2}{\pi} U_{ttx} - \frac{\partial}{\partial x} \left(\frac{\mu \nu^2 S^2}{\pi} U_{xx} \right) \right], \qquad (2.27)$$

where the Lame coefficient: $\mu = E/(2(1+\nu))$ was introduced.

Now it is possible to rewrite it for formal analysis in terms of the strain component $u = U_x$, that will be derived by differentiation as:

$$u_{tt} - \frac{\partial}{\partial x} \left[\frac{1}{\rho S} (ESu)_x \right] = \frac{\partial}{\partial x} \left[\frac{1}{2\rho S} \frac{\partial}{\partial x} \left[S\beta u^2 + \frac{\rho \nu^2 S^2}{\pi} u_{tt} - \frac{\partial}{\partial x} \left(\frac{\mu \nu^2 S^2}{\pi} u_x \right) \right] \right]. \qquad (2.28)$$

Two particular cases are of interest.

When the rod is *inhomogeneous* and uniform $R(x) = R_0 - const$, the equation (2.27) gives:

$$U_{tt} - \frac{1}{\rho} (EU_x)_x = \frac{1}{2\rho} \frac{\partial}{\partial x} \left[\beta U_x^2 + \rho \nu^2 R_0^2 U_{ttx} - R_0^2 \frac{\partial}{\partial x} \left(\mu \nu^2 U_{xx} \right) \right], \qquad (2.29)$$

and in terms of the strain component $u = U_x$ we have :

$$u_{tt} - \frac{\partial}{\partial x} \left[\frac{1}{\rho} (Eu)_x \right] = \frac{\partial}{\partial x} \frac{1}{2\rho} \frac{\partial}{\partial x} \left[\beta u^2 + \rho \nu^2 R_0^2 u_{tt} - R_0^2 \frac{\partial}{\partial x} \left(\mu \nu^2 u_x \right) \right]. \qquad (2.30)$$

Another reduction appears when the rod is *non-uniform* but the homogeneous one:

$$U_{tt} - \frac{c_0^2}{R^2} \frac{\partial}{\partial x} \left(R^2 U_x \right) = \frac{1}{2R^2} \frac{\partial}{\partial x} \left[\frac{\beta}{\rho} R^2 U_x^2 + \nu^2 R^4 U_{ttx} - c_1^2 \nu^2 \frac{\partial}{\partial x} \left(R^4 U_{xx} \right) \right], \qquad (2.31)$$

where $c_0^2 = E/\rho$ and $c_1^2 = \mu/\rho$ are the linear longitudinal and shear wave velocities, respectively. Writing it via u, we have

$$u_{tt} - c_0^2 \frac{\partial}{\partial x}\left[\frac{1}{R^2}\frac{\partial}{\partial x}\left(R^2 u\right)\right] = \frac{\partial}{\partial x}\frac{1}{2R^2}\frac{\partial}{\partial x}\left[\frac{\beta}{\rho}R^2 u^2 + \nu^2 R^4 u_{tt} - c_1^2 \nu^2 \frac{\partial}{\partial x}\left(R^4 u_x\right)\right].$$

In the simplest case of uniform and homogeneous rod we have for the longitudinal displacement U the Doubly Dispersive Equation, the DDE

$$U_{tt} - c_0^2 U_{xx} = \frac{\beta}{2\rho}(U_x^2)_x + \frac{\nu^2 R^2}{2}(U_{tt} - c_1^2 U_{xx})_{xx} \tag{2.32}$$

as well as in terms of the strain component $u = U_x$:

$$u_{tt} - c_0^2 u_{xx} = \left[\frac{\beta}{2\rho}u^2 + \frac{\nu^2 R^2}{2}(u_{tt} - c_1^2 u_{xx})\right]_{xx}, \tag{2.33}$$

where all coefficients possess the dimensional values still, and both u and β were assumed to be positive. Both (2.32) and (2.33) were obtained using the Lagrange formalism, perhaps, firstly in (Samsonov, 1982).

To obtain the dimensionless version of an equation, we introduce the scale F for a variable f such as for any f the dimensionless quantity is defined as $\overline{f} = f/F$, e.g., the derivative in x as $\partial/\partial x = (1/X)\partial/\partial\overline{x}$. In addition, we assume that the problem will be considered only for a wave, propagating with a velocity close to the linear longitudinal ('rod's') wave velocity c_0^2. It allows the ability to introduce the necessary relation between the scales for time T, space variable X and the linear wave velocity c_0^2 as follows: $X^2 = T^2 E/\rho$. We have:

$$u_{tt} - \frac{\partial}{\partial x}\left[\frac{1}{\rho}\frac{\partial}{\partial x}(Eu)\right] = \frac{1}{2}\frac{\partial}{\partial x}\left[\frac{1}{\rho}\frac{\partial}{\partial x}\left(\frac{BA}{E}\beta u^2 + \frac{R^2 N^2}{X^2}\rho\nu^2 u_{tt} - \frac{\partial}{\partial x}\left(\frac{MN^2 R^2}{EX^2}\mu\nu^2 u_x\right)\right)\right],$$

where bars are omitted for brevity, A is the scale for the strain magnitude, and the capital letters B, M, N are used for corresponding scale values for β, μ, ν. Further reduction is based on the assumptions that

characteristic strain magnitude is *small* enough, $A < 1$

long waves are considered only: $R^2/X^2 << 1$

the balance relation for nonlinear and dispersive terms is valid: $A = R^2/X^2$,

then the introduction of the small parameter:

$$\varepsilon = A = \frac{R^2}{X^2} \tag{2.34}$$

yields the dimensionless DDE for the wave propagation in the inhomogeneous circular cylinder under the assumptions described above:

$$u_{tt} - \frac{\partial}{\partial x}\left[\frac{1}{\rho}\frac{\partial}{\partial x}(Eu)\right] = \frac{\varepsilon}{2}\frac{\partial}{\partial x}\left[\frac{1}{\rho}\frac{\partial}{\partial x}\left(p\beta u^2 + \rho\nu^2 u_{tt} - \frac{\partial}{\partial x}\left(b\mu\nu^2 u_x\right)\right)\right] \tag{2.35}$$

where $b = M/E < 1$ and $p = B/E$ are the combinations of constant scale factors.

2.3.2 Refinement of the derivation of the DDE

Two things are to be mentioned here.

First, it would be useful to confirm the Love hypothesis formally, second, it provides commonly used but not the unique relationship between V and U. In the last decade, this model has been widely applied to the nonlinear strain waves description in rods, but the limit of it is that the boundary conditions on a free lateral surface were not properly taken into account, when these hypotheses were formulated. Direct substitution of these assumptions into the conditions of absence of both normal and tangential stresses at the lateral surface of a free nonlinear rod does not necesarily result in the zero stresses. Generally speaking, the identity is not required because an asymptotic solution is to be found, however the boundary conditions failure may indicate possible neglecting of several terms of the same order that cannot be recovered from these hypotheses.

Secondly, the Poisson effect (2.21) results from physics, and the question arises on how to calculate a next term in the dependence of V of U_x, if any.

For the wave propagation problem in a uniform free rod the components P_{rr}, P_{rx} of the Piola-Kirchhoff stress tensor $\mathbf{P}$, defined in reference configuration, should vanish at the lateral surface $r = R_0$ free of normal and tangential stresses. For the 5-constant (Murnaghan) nonlinear elasticity, involving the moduli $(\lambda,\ \mu,\ l,\ m,\ n)$ they are:

$$
\begin{aligned}
P_{rr} = {} & (\lambda + 2\mu)V_r + \lambda\frac{V}{r} + \lambda U_x + \\
& \frac{\lambda + 2l}{2}\left(\frac{V^2}{r^2} + 2V_r\frac{V}{r}\right) + \frac{\lambda + 2\mu + m}{2}U_r^2 + \frac{3\lambda + 6\mu + 2l + 4m}{2}V_r^2 + \\
& (\mu + m)U_r V_x + (\lambda + 2l)U_x V_r + (2l - 2m + n)U_x\frac{V}{r} + \\
& \frac{\lambda + 2l}{2}U_x^2 + \frac{\lambda + 2\mu + m}{2}V_x^2,
\end{aligned}
\tag{2.36}
$$

$$
\begin{aligned}
P_{rx} = {} & \mu U_r + \mu V_x + \\
& (\lambda + 2\mu + m)U_r V_r + \frac{2\lambda + 2m - n}{2}U_r\frac{V}{r} + \\
& \frac{2m - n}{2}V_x\frac{V}{r} + (\mu + m)V_x V_r + (\lambda + 2\mu + m)U_x U_r + \\
& (\mu + m)U_x V_x.
\end{aligned}
\tag{2.37}
$$

To satisfy the zero stess condition we have to refine some of assumptions *i)-iv)*. We will consider possible generalization of the relationship $V(U_x)$, given by the Love hypothesis as the most restrictive one. We introduce the small parameter ε, taking into account that the strain waves under study should be the *elastic* waves with sufficiently small magnitude $A \ll 1$, as well as the sufficiently long waves with the length Λ, so as the ratio $R_0/\Lambda \ll 1$, will be valid, where R is defined by $r \le R_0$ in the uniform rod. The most important case occurs when both nonlinear and dispersive features are *in balance* and *small enough:*

$$\varepsilon = A = \left(\frac{R_0}{\Lambda}\right)^2 \ll 1. \tag{2.38}$$

We introduce $\tilde{U} \equiv A\Lambda$ as the scale value for displacements U and V, and Λ as the scale for the distances along the rod $x-$ axis. Then the boundary conditions of the absence of stresses at the free lateral surface $P_{rr} = 0$, $P_{rx} = 0$ can be written in the dimensionless form after (2.36),(2.37) in power series of ε:

$$(\lambda + 2\mu)V_r + \lambda\frac{V}{R} + \lambda U_x + \frac{\lambda + 2\mu + m}{2}U_r^2 +$$

$$\varepsilon\left[\frac{3\lambda + 6\mu + 2l + 4m}{2}V_r^2 + \frac{(\lambda + 2l)}{2}\left(\frac{2VV_r}{R} + \frac{V^2}{R^2} + 2U_xV_r\right) + \right.$$

$$\left.(2l - 2m + n)\frac{VU_x}{R} + (\mu + m)U_rV_x + \frac{\lambda + 2l}{2}U_x^2\right] +$$

$$\varepsilon^2\left(\frac{\lambda + 2\mu + m}{2}V_x^2\right) = 0, \tag{2.39}$$

and

$$\mu U_r +$$

$$\varepsilon\left(\mu V_x + (\lambda + 2\mu + m)U_rV_r + (2\lambda + 2m - n)U_r\frac{V}{R}\right) +$$

$$\varepsilon^2\left((\lambda + 2\mu + m)U_xU_r + \left(\frac{2m - n}{2}\right)\frac{VV_x}{R} + (\mu + m)V_xV_r\right) +$$

$$\varepsilon^3(\mu + m)U_xV_x = 0. \tag{2.40}$$

We expand the unknown functions U, V in power series of ε:

$$U = U_0 + \varepsilon U_1 + \varepsilon^2 U_2 + \ldots, \quad V = V_0 + \varepsilon V_1 + \varepsilon^2 V_2 + \ldots. \tag{2.41}$$

then substituting (2.41) in (2.39), (2.40), and equating to zero all terms of the same order of ε. One can conclude that the plane cross section hypothesis and the Love hypothesis are valid in the leading order only:

$$U_0 = U(x,t), \quad V_0 = -\frac{\lambda}{2(\lambda + \mu)}rU_x \equiv -\nu rU_x. \tag{2.42}$$

Terms proportional to $O(1)$ confirm their asymptotical validity now. Next order terms yield:

$$U_1 = \frac{\nu}{2}r^2U_{xx}, \tag{2.43}$$

$$V_1 = -\frac{\nu^2}{2(3 - 2\nu)}r^3U_{xxx} -$$

$$\left[\frac{\nu(1 + \nu)}{2} + \frac{(1 - 2\nu)(1 + \nu)}{E}\left(l(1 - 2\nu)^2 + 2m(1 + \nu) - n\nu\right)\right]rU_x^2,$$

$$\tag{2.44}$$

where ν is the Poisson coefficient and E is the Young modulus. Other terms from the series in (2.41) for $i > 2$ may be found in the same way, however they will be omitted here because of negligible influence on the final model equation for the strain waves.

The expansion in ε can be written in the linear elasticity case as:

$$U = U(x,t) + \varepsilon\,\frac{\nu}{2}r^2 U_{xx} + O(\varepsilon^2); \tag{2.45}$$

$$V = -\nu r U_x - \varepsilon\frac{\nu^2}{2(3-2\nu)}r^3 U_{xxx} + O(\varepsilon^2) \tag{2.46}$$

which is sufficient for further asymptotic analysis.

Therefore we began with the asymptotic expansion of V and U with the power series in ε, using the assumptions *i)-iv)* about the type of wave and obtained the relationship between V and U as the power series with respect to the radial co-ordinate r. In fact, both equations $P_{rr} = 0$, $P_{rx} = 0$ are valid on the lateral surface $r = R$, and not necessarily inside a rod. The solutions of the equations followed from the boundary conditions written in (2.39) and (2.40) are not unique.

Hence we can *assume* now that both displacements can be expanded into power series with respect to r, and it is quite instructive to apply the power series in r in order to get a relation $V(U_x)$ as reasonable extrapolation of the dependence valid at the lateral surface. The small strain amplitude condition $A \ll 1$ allows to consider separately linear and nonlinear terms in U and V, while the expansion in power series of r will be assumed in the long wave limit. It yields:

$$U = U^L + U^{NL}; \tag{2.47}$$
$$U^L = U_0(x,t) + r\,U_1(x,t) + r^2 U_2(x,t) + .., $$
$$U^{NL} = U_0^{NL}(x,t) + r U_1^{NL}(x,t) + .., \tag{2.48}$$

$$V = V^L + V^{NL}; \tag{2.49}$$
$$V^L = V_0(x,t) + r\,V_1(x,t) + r^2 V_2(x,t) + .., $$
$$V^{NL} = V_0^{NL}(x,t) + r\,V_1^{NL}(x,t) + .. \tag{2.50}$$

Substituting (2.47), (2.49) into $P_{rr} = 0$, $P_{rx} = 0$ for $r = R$ and collecting terms with equal power of r, we get the dependence of r in the form of the same quantities as above:

$$U = U(x,t) + \nu r^2 U_{xx} + ..., \tag{2.51}$$

$$V = -\nu r U_x - \frac{\nu^2}{2(3-2\nu)}r^3 U_{xxx} - $$
$$\left(\frac{\nu(1+\nu)}{2} + \frac{(1-2\nu)(1+\nu)}{E}\left[l(1-2\nu)^2 + 2m(1+\nu) - n\nu\right]\right)rU_x^2 + ..., $$
$$\tag{2.52}$$

and, similarly, the higher order terms may be calculated.

Moreover, the 1st term (depending on linear elasticity) in V_1 is enough for vanishing all the components of the Piola-Kirchhoff stress tensor on the lateral surface with prescribed accuracy.

We will refer to the following formula

$$V = -\nu r U_x - \frac{\nu^2}{2(3-2\nu)} r^3 U_{xxx} \tag{2.53}$$

as the refined relationship for $V(U_x)$.

Now we may derive the refined version of the DDE in the dimensional form.

To obtain it, the Lagrangian may be written without higher order nonlinear and differential terms in the relationships for kinetic energy K and potential strain energy Π. Substituting expansions (2.51,2.53) into the formulae for K and Π, written for the Murnaghan material, one can find respectively:

$$K = \frac{\rho_0}{2} \left(U_t^2 + \nu r^2 \left[U_t U_{xxt} + \nu U_{xt}^2 \right] \right), \tag{2.54}$$

$$\Pi = \frac{1}{2} \left(E U_x^2 + \frac{\beta}{3} U_x^3 + \nu E r^2 U_x U_{xxx} \right). \tag{2.55}$$

It is easy to see that the usage of truncated expansion for V, containing only two first terms, is sufficient to write expressions for K and Π with prescribed accuracy. Substituting them into (2.1) and calculating $\delta S = 0$, one can obtain the *dimensional* refined DDE, useful for applications to physical experiments:

$$U_{tt} - c_0^2 U_{xx} = \frac{\beta}{2\rho} (U_x^2)_x - \frac{\nu(1-\nu)R^2}{2} U_{ttxx} + \frac{\nu R^2 c_0^2}{2} U_{xxxx}, \tag{2.56}$$

The coefficients with the dispersive terms here differ from those in (2.32) due to the terms U_1, V_1, resulted after the boundary conditions fulfillment at the free lateral surface, as it was done in (Porubov, Samsonov, 1993). The difference with the simplest analysis made above is that now both assumptions *i)* and *ii)* are confirmed using asymptotics. In terms of the component of the linear longitudinal strain $u = U_x$ the refined DDE is written as:

$$u_{tt} - c_0^2 u_{xx} = \left[\frac{\beta}{2\rho} u^2 - \frac{\nu(1-\nu)R^2}{2} u_{tt} + \frac{\nu R^2}{2} c_0^2 u_{xx} \right]_{xx}, \tag{2.57}$$

The only difference between (2.33) and (2.57) is in values of coefficients with dispersive terms, not very important for formal analysis. We have taken into account formally in (2.57) possible non-planar cross section. Surprisingly, the estimation of a cross section curvature for real materials results in the value of several percent, (Porubov, Samsonov, 1993), nevertheless it is enough to avoid any formal discrepancy in free lateral surface condition.

What counts is the main difference between any version of the DDE, say, (2.33), (2.57) and the well known Boussinesq equation derived for shallow water waves, consisting in simultaneous and competitive action of two different dispersive terms. This is the distinctive feature of wave phenomena in solids, where shear waves exist, are of the same order as the longitudinal waves, and should be taken into consideration together with them. Any longitudinal motion inside solids causes immediate reaction in transversal direction because of crystalline or grain structure. A liquid does not have shape due to negligible shears.

Moreover, in contrast to the shallow water problem, both β and u in the nonlinear guided wave problem for elastic solids can be of different signs.

Aiming to derive the *dimensionless* version of the DDE for strain u we write explicitly $u = |u|$ sgn u and $\beta = |\beta|$ sgn β, where: $\forall z$ sgn $z = +1$ for $z > 0$ and sgn $z = -1$ for $z < 0$. Introducing the wave length Λ as the scale, the dimensionless variable : $\xi = x/\Lambda$, and the time scale as $T = \Lambda/c_0$, that leads to the dimensionless time variable $\tau = t/T$, we obtain from (2.57) for $|u| = A f(\xi, \tau)$

$$f_{\tau\tau} - f_{\xi\xi} = \frac{1}{2}\left[A\frac{|\beta|}{E}f^2 \text{sgn}\beta \text{ sgn}u - \frac{\nu(1-\nu)R^2}{L^2}f_{\xi\xi} + \frac{\nu R^2}{L^2}f_{\tau\tau}\right]_{\xi\xi}.$$

To find a stable solitary or periodical nonlinear wave solution we consider waves, which length provides the balance of nonlinearity and dispersion, and obtain the dimensionless DDE for the strain[1] function:

$$f_{\tau\tau} - f_{\xi\xi} = \varepsilon\left[pf^2 + af_{\tau\tau} + bf_{\xi\xi}\right]_{\xi\xi},$$

$$p = \frac{|\beta|}{E}\text{sgn}\beta \text{ sgn}u; \quad a = \nu; \quad b = -\nu(1-\nu)$$

The standard shift transformation yields the following dimensionless equation for further formal analysis in Chapter 3:

$$f_{\tau\tau} - f_{\xi\xi} = \left[pf^2 + af_{\tau\tau} + bf_{\xi\xi}\right]_{\xi\xi} \qquad (2.58)$$

For $b = 0$ the Boussinesq equation results from it, while for $a = 0$ it can be reduced to the Zakharov equation governing the nonlinear string problem as a continuum limit of a nonlinear lattice oscillations problem, see also (Clarkson, LeVeque, Saxton, 1986) and (Kunin, 1975; 1982), respectively. The corresponding Boussinesq limit of the DDE in solid state physics is often called the nonlinear acoustic limit.

As a rule, wave solutions of (2.58) differ from those of evolution equations of the KdV-type and of the Boussinesq equation. Note that two dispersive terms in the DDE may formally compensate each other. Below we will show how important is the shear influence in experiments in solitons in solids.

Even in the corresponding linearized problem one can reveal a difference in dispersion properties of the KdV equation, the DDE and the Boussinesq equations, when either f_{ttxx} or f_{xxxx} is taken into account. As one can see in the Figures in Chapter 1, when dispersive term is proportional to the fourth derivative in space, the linear wave frequency ω in the corresponding dispersion relation tends to infinity for wave number $k \to \infty$. On the contrary, when dispersion is proportional to the fourth order mixed derivative, there is a finite limit for ω when $k \to \infty$. The linearized version of the

---

[1]Evidently, the strain is dimensionless, while the equation (2.57) was not.

DDE provides a sector n the (ω, k) plane, in which the dispersion curve is localized, and $\omega(k)$ tends to $\omega = k$ for $k \to \infty$.

It should be noted that the nonlinear equations (2.32), (2.58) are quasi-hyperbolic ; that is the reason why the complete set of boundary and initial conditions for any wave problem can be satisfied, in contrast to the corresponding evolution equation, which is actually a result of widespread, but not always well-grounded, reduction of (2.58).

2.3.3 Equations for wave in non-uniform highly nonlinear wave guide

The same approach is used for the derivation of the nonlinear wave propagation equation for the displacement U in the framework of the so called "nine-constants theory" for Murnaghan's material. It has the following form:

$$
\begin{aligned}
U_{tt} - c_0^2 U_{xx} &= \frac{\beta}{2\rho}(U_x^2)_x - \frac{\nu(1-\nu)R^2}{2}U_{ttxx} + \frac{\nu R^2 c_0^2}{2}U_{xxxx} \\
&+ \frac{4a}{\rho}(U_x^3)_{xxx} - \frac{bR^2}{2\rho}\left[U_{xx}^2 - \frac{1}{2}(U_x^2)_{xx}\right]_x,
\end{aligned} \tag{2.59}
$$

where the constants depend on the Murnaghan elastic moduli of the 3d (l, m, n) and 4th order $(\nu_1, \nu_2, \nu_3, \nu_4)$, see Chapter 1, and the coefficients are:

$$
\begin{aligned}
a &= \frac{E(3\lambda^3 + 6\lambda^2\mu + 8\lambda\mu^2 + 4\mu^3)}{32\mu(\lambda+\mu)^2} + \frac{l}{2}(1-2\nu)^2(1+2\nu^2) + \frac{n\nu^2(1-2\nu)}{2} \\
&+ m(1+\nu)^2(1-2\nu+2\nu^2) + \\
&\frac{\mu^4}{(\lambda+\mu)^4}\nu_1 - \frac{\lambda\mu^2(3\lambda+4\mu)}{4(\lambda+\mu)^4}\nu_2 + \nu^2(1-2\nu)\nu_3 + \nu^2(4-4\nu+\nu^2)\nu_4; (2.60)
\end{aligned}
$$

$$
b = \frac{n\nu^3}{4} + \frac{m\nu^2}{2}(1-2\nu) + \nu^2\mu \tag{2.61}
$$

Introduction of the strain yields the nonlinear strain wave equation for u in the framework of the 'nine-constants theory' :

$$
\begin{aligned}
u_{tt} - c_0^2 u_{xx} &= \frac{1}{2}\left[\frac{\beta}{\rho}u^2 - \nu(1-\nu)R^2 u_{tt} + \nu R^2 c_0^2 u_{xx} + \right. \\
&\left. \frac{8a}{\rho}(u^3)_{xx} - \frac{bR^2}{\rho}(u_x^2 - \frac{1}{2}(u^2)_{xx})\right]_{xx}
\end{aligned} \tag{2.62}
$$

whilst the transversal displacement was defined by means of the refined relation (2.53).

2.4   Main assumptions and 2-D coupled equations

Natural generalisation of the 1+1D nonlinear wave propagation problem in solids consists in the problem for the two-dimensional wave guides, e.g., for a thin nonlinearly elastic plate. The plate is assumed to be semi-infinite in two directions x and y, and very thin in z. A wave under consideration is supposed to propagate in a direction coplanar to the median plane of the plate.

Consider the problem of longitudinal deformation waves propagation in the homogeneous isotropic infinite nonlinear elastic plate. We introduce the Lagrangian cartesian coordinates $\mathbf{x}(x, y, z)$, such that the median plane of the plate is described by the equation $z = 0$, while the lateral planes are described by the equations $z = \pm h/2$, where h is the thickness of the plate. The derivation of the system of equations is based also on Hamilton's variational principle and on the assumption that the deformation of the material is described by relationships of the nonlinear theory of elasticity of either incompressible or compressible medium.

In general, the tensor $\mathbf{C}$ components can be written as :

$$
\begin{aligned}
2C_{xx} &= (1 + U_x)^2 - 1 + V_x^2 + W_x^2 & C_{xy} &= C_{yx} \\
2C_{yx} &= V_x + U_y + U_x U_y + V_x V_y + W_x W_y & 2C_{yy} &= (1 + V_y)^2 - 1 + U_y^2 + W_y^2 \\
2C_{zx} &= W_x(1 + W_z) & 2C_{zy} &= W_y(1 + W_z)
\end{aligned}
$$

$$
\begin{aligned}
C_{xz} &= C_{zx} \\
C_{yz} &= C_{zy} \\
2C_{zz} &= (1 + W_z)^2 - 1
\end{aligned}
\tag{2.63}
$$

which will hardly allow to decrease the dimension of an initial 3D problem.

In the rod problem the simplifications were based on either Love's postulate or on the refined formula between V and U_x, therefore it seems to be reasonable to derive similar relations in the 2D case. Again, there is the difference in reduction of equations for incompressible and for compressible materials: dealing with the first, it is enough to use directly the additional incompressibility condition, whilst in the second case one has to install a relation of U and V.

If the deformations are symmetric with respect to the plane $z = 0$ of the plate, while the characteristic wave length greatly exceed h, the displacements $\mathbf{U}(U, V, W)$ are approximately calculated as follows (Grigolyuk, Selezov, 1973): $U = U(x, y, t)$, $V = V(x, y, t)$, $W = zW(x, y, t)$, where t is time.

For a wave propagation problem in the *incompressible* plate, the relationship under consideration can be derived rigorously. Indeed, taking into account the condition of incompressibility $\det |\mathbf{G}| = 1$, where $\mathbf{G}$ is the Cauchy-Green tensor of measure of deformation (see Chapter1), it enables one to find the explicit expression for the displacement W perpendicular to the midplane of the plate in terms of the in-plane displacements U and V:

$$
W = z\{[(1 + U_x)(1 + V_y) - U_y V_x]^{-1} - 1\}, \tag{2.64}
$$

that allows to reduce the problem to the 2D one. Indeed, e.g., evaluating the Lagrangian density $\mathcal{L}$ for the incompressible Mooney material:

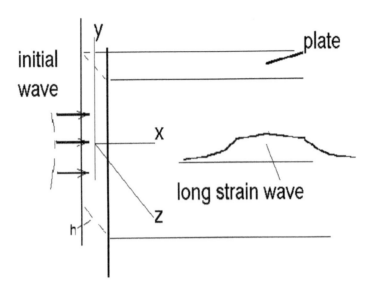

Figure 2.1: Geometry of wave problem in a plate.

$$\mathcal{L} = \rho \mathbf{u}_t^2/2 - \mu/4\left[(1+l)(I_1(\mathbf{G})-3) + (1-l)(I_2(\mathbf{G})-3)\right], \qquad (2.65)$$

where ρ is the density of the material, μ is the shear modulus, l is the only third order modulus of elasticity, $|l| \le 1$, and I_k are the invariants of $\mathbf{G}$, and writing down the conditions that the action functional possess the stationary value, we obtain the following coupled highly nonlinear equations for waves in the Mooney plate:

$$\begin{cases} U_{tt} - U_{xx} - (1/4)(U_{yy} + 3V_{xy}) = \varepsilon\left[(1/48)(4s_{tt} - \Delta s)_x + P_x + Q_y\right] \\[2mm] V_{tt} - V_{yy} - (1/4)(V_{xx} + 3U_{xy}) = \varepsilon\left[(1/48)(4s_{tt} - \Delta s)_y + P_y^1 + Q_x^1\right] \end{cases} \qquad (2.66)$$

where the displacement gradient component is introduced:

$$s \equiv (U_x + V_y)$$

and

$$\begin{aligned} 4P \equiv{}& -\left[6U_x^2 + V_y^2 + (1/2)(U_y^2 + V_x^2) + 4U_y V_x + 2U_x V_y\right] + \\ & (l/2)\left[(U_y + V_x)^2 - 4V_y(2U_x + V_y)\right], \end{aligned}$$

$$4Q \equiv s\left[l(U_y + V_x) - U_y - 4V_x\right].$$

The values P^1, Q^1 used in the second equation can be written similarly to the expressions for P, Q after formal replacement of x by y, y by x, U by V, and V by U. To derive the equation (2.66) we have introduced, as before, the dimensionless variables according to the formal rule $\overline{f} = f/F$, where F is the scale value for the dimensional quantity f. The small parameter ε, that will be used for solution of the equations, has been given in such a way that the terms describing the influence of the nonlinearity of the material and the dispersion of waves in $\mathcal{L}$ should be of the same order of magnitude and small enough in comparison with kinetic and potential energy of the plate calculated within the framework of the linear theory:

$$\varepsilon = A/X = B/Y = C/Z = (Z/X)^2 << 1$$

for $Z = \pm h/2$; quantities A, B, C represent the scales for displacements (U, V, W) respectively. When the effect of nonlinearity is not taken into account $(P = Q = 0)$ the system (2.66) is reduced to the well known equations of the refined theory of longitudinal waves in the incompressible plates for the so-called two-mode model of symmetric motions in a thin layer, see (Grigolyuk and Selezov, 1973).

The derivation of similar reductions of the initial 3D problem to the 2D one for the Murnaghan compressible nonlinearly elastic material is different. It could be done using the approximation based on the solution of the generalized plane-stressed state problem:

$$W = -\frac{\nu z}{1 - \nu}(U_x + V_y)$$

where ν is the Poisson ratio, as it was assumed in (Grigolyuk, Selezov, 1973).

We will derive the formula using the approach similar to that applied in rod's problem. On the free lateral surface of the plate all components of the Piola-Kirchhoff stress tensor must vanish:

$$P_{zz} = 0; P_{zx} = 0; P_{zy} = 0.$$

Explicit expressions for each of them are cumbersome, e.g., the component P_{zz} may be written in displacements as:

$$
\begin{aligned}
P_{zz} = & \ \lambda U_x + (\frac{1}{2}\lambda + l)U_x^2 + \frac{1}{2}(\lambda + m - \frac{1}{2}n)U_y^2 + \frac{1}{2}(\lambda + 2\mu + m)U_z^2 + \frac{l + 2m}{4}U_z^4 + \\
& (2l - 2m + n)U_x V_y + (l + m)U_x U_z^2 + lU_x V_z^2 + (m - \frac{1}{2}n)U_y V_x + lU_z^2 V_y + \\
& (\mu + m)U_z W_x + (\frac{1}{2}\lambda + \mu + l + \frac{5}{2}m)U_z^2 W_z + (\lambda + 2l)U_x W_z + \lambda V_y + \\
& m(U_y U_z V_z + U_z V_x V_z) + (\frac{1}{2}l + m)U_z^2 V_z^2 + \frac{2\lambda + 2m - n}{2}V_x^2 + (\frac{1}{2}\lambda + l)V_y^2 + \\
& \frac{l + 2m}{4}V_z^4 + (l + m)V_y V_z^2 + \frac{\lambda + 2\mu + m}{2}V_z^2 + (\mu + m)V_z W_y + (\lambda + 2\mu)W_z + \\
& (\frac{1}{2}\lambda + \mu + l + \frac{5}{2}m)V_z^2 W_z + (\lambda + 2l)V_y W_z + (\frac{3}{2}\lambda + 3\mu + l + 2m)W_z^2.
\end{aligned}
$$

We apply the asymptotic expansion method to simplify the problem. The small parameter ε will be introduced, taking into account that longitudinal waves of displacements (U, V, W) in the (x, y) plane should be *elastic* waves of sufficiently small magnitudes $A, B, C << 1$, as well as sufficiently long waves. The most important case occurs when both nonlinear and dispersive features are *in balance* and *small enough*. We introduce $\tilde{U} \equiv AX$ as the scale value for U and $\tilde{V} \equiv BY$, and X, Y as the scales for distances along the corresponding axes in a plane perpendicular to the $z-$ axis. Eventually the small parameter is:

$$\frac{A}{h} = \frac{B}{h} = \varepsilon$$

$$\varepsilon^2 = \frac{A}{X} = \frac{B}{Y} = \frac{B}{X} = \frac{B}{Y} = \frac{C}{h} = \left(\frac{h}{X}\right)^2 = \left(\frac{h}{Y}\right)^2 = \frac{h^2}{XY}$$

$$\frac{C}{X} = \frac{C}{Y} = \varepsilon^3.$$

In dimensionless form the conditions of vanishing of all components of the Piola-Kirchoff stress tensor on the surface of the plate yield the following truncated expressions:

$$
\begin{aligned}
P_{zz} &= [\lambda(U_x + V_y) + \frac{1}{2}(\lambda + 2\mu + m)(U_z^2 + V_z^2) + (\lambda + 2\mu)W_z] + \\
&\quad \varepsilon^2[(\frac{1}{2}\lambda + l)U_x^2 + (l + m)U_xU_z^2 + \frac{1}{2}(m + \frac{1}{2}l)(U_z^2 + V_z^2)^2 + \frac{\lambda + m - n}{4}U_y^2 + \\
&\quad mU_yU_zV_z + mU_zV_xV_z + lU_z^2V_y + (\frac{1}{2}\lambda + \mu + l + \frac{5}{2}m)U_z^2W_z + \\
&\quad (2l - 2m + n)U_xV_y + (m - \frac{1}{2}n)U_yV_x + (\mu + m)U_zW_x + (\lambda + 2l)U_xW_z + \\
&\quad \frac{1}{2}(\lambda + m - \frac{1}{2}n)V_x^2 + (\frac{1}{2}\lambda + l)V_y^2 + (l + m)V_yV_z^2 + (\mu + m)V_zW_y + lU_xV_z^2 + \\
&\quad (\frac{1}{2}\lambda + \mu + l + \frac{5}{2}m)V_z^2W_z + (\lambda + 2l)V_yW_z + (\frac{3}{2}\lambda + 3\mu + l + 2m)W_z^2] \\
&= 0
\end{aligned} \tag{2.67}
$$

$$
\begin{aligned}
P_{zx} &= \mu U_z + \\
&\quad \varepsilon^2[(\frac{1}{2}\lambda + \mu + m)U_z^3 + (\lambda + 2\mu + m)U_xU_z + (\lambda + m - \frac{1}{2}n)U_zV_y + \\
&\quad (\mu + \frac{1}{4}n)U_yV + \frac{1}{2}(\lambda + 2\mu + m)U_zV_z^2 + (\lambda + 2\mu + m)U_zW_z + (\mu + \frac{1}{4}n)V_xV_z \\
&\quad + \mu W_x] + \\
&\quad \varepsilon^4[(\mu + m)U_xW_x + (m - \frac{1}{2}n)V_yW_x + (\mu + \frac{1}{4}n)U_yW_y + \frac{1}{4}nV_xW_y + \\
&\quad (\mu + m)W_xW_z + ...] = 0
\end{aligned} \tag{2.68}
$$

$$
\begin{aligned}
P_{zy} &= \mu V_z + \\
&\quad \varepsilon^2[(\mu + \frac{1}{4}n)U_yU_z + (\lambda + m - \frac{1}{2}n)U_xV_z + (\mu + \frac{1}{4}n)U_zV_x + \\
&\quad \frac{1}{2}(\lambda + 2\mu + m)U_z^2V + (\frac{1}{2}\lambda + \mu + m)V_z^3 + (\lambda + 2\mu + m)V_yV_z + \\
&\quad (\lambda + 2\mu + m)V_zW_z + \mu W_y] + \\
&\quad \varepsilon^4[(m - \frac{1}{2}n)U_xW_y + \frac{1}{4}nU_yW_x + (\mu + \frac{1}{4}n)V_xW_x + (\mu + m)V_yW_y + \\
&\quad (\mu + m)W_yW_z + ...] = 0
\end{aligned} \tag{2.69}
$$

Assuming the power series expansions for displacements:

$$
\begin{aligned}
U &= U_0 + \varepsilon U_1 + ...; \\
V &= V_0 + \varepsilon V_1 + ...; \\
W &= W_0 + \varepsilon W_1 + ...
\end{aligned}
\tag{2.70}
$$

we obtain in order of ε^0 :

$$
\begin{aligned}
\lambda(U_{0,x} + V_{0,y}) + (\lambda + 2\mu)W_{0,z} &= 0, \\
\mu U_{0,z} &= 0, \\
\mu V_{0,z} &= 0,
\end{aligned}
$$

hence the leading order terms in displacements are independent of z:

$$
U_0 = U_0(x, y; t); \quad V_0 = V_0(x, y; t),
$$

and the most important result is that :

$$
W_{0,z} = -\frac{\lambda}{\lambda + 2\mu}(U_{0,x} + V_{0,y}) = -\frac{\nu}{1 - \nu}(U_{0,x} + V_{0,y}),
$$

and:

$$
W_0 = -\frac{\nu z}{1 - \nu}(U_{0,x} + V_{0,y}),
\tag{2.71}
$$

that means that the relation between transversal displacement and in-plane strains is confirmed to be equivalent to the generalized plane stress problem solution: W_0 is proportional to z and to the functions of x and y.

To get the next order terms (refinements) in (2.70) we have to consider the problem in $O(\varepsilon^1)$:

$$
\begin{aligned}
\lambda(U_{1,x} + V_{1,y}) + (\lambda + 2\mu)W_{1,z} &= 0, \\
\mu U_{1,z} &= 0, \\
\mu V_{1,z} &= 0,
\end{aligned}
$$

that result in:

$$
U_1 = 0, \quad V_1 = 0, \quad W_1 = W(x, y; t),
$$

where $W(x, y; t)$ is an unknown function, independent of z.

The second order problem is reduced to the following equations:

$$
\begin{aligned}
0 = {} & \lambda(U_{2,x} + V_{2,y}) + (\lambda + 2\mu)W_{2,z} + (\tfrac{1}{2}\lambda + l)U_x^2 + \tfrac{1}{2}(\lambda + m - \tfrac{1}{2}n)U_y^2 + \\
& (2l - 2m + n)U_x V_y + (m - \tfrac{1}{2}n)U_y V_x + (\lambda + 2l)U_x W_{0,z} + (\tfrac{1}{2}\lambda + l)V_y^2 + \\
& \tfrac{1}{2}(\lambda + m - \tfrac{1}{2}n)V_x^2 + (\lambda + 2l)V_y W_{0,z} + (\tfrac{3}{2}\lambda + 3\mu + l + 2m)W_{0,z}^2; \\
0 = {} & \mu U_{2,z} + \mu W_{0,x}; \\
0 = {} & \mu V_{2,z} + \mu W_{0,y};
\end{aligned}
$$

Two last equations yield:

$$U_2 = \frac{\nu z^2}{2(1-\nu)}(U_{0,xx} + V_{0,yx}),$$

$$V_2 = \frac{\nu z^2}{2(1-\nu)}(U_{0,xy} + V_{0,yy}),$$

$$W_2 = -\frac{z}{(\lambda+2\mu)}\left[\left(\frac{\lambda(\lambda+\mu)}{(\lambda+2\mu)} + \frac{4l\mu^2}{(\lambda+2\mu)^2} + \frac{2m\lambda^2}{(\lambda+2\mu)^2}\right)U_{0,x}^2 + \right.$$
$$\frac{1}{2}\left(\lambda+m-\frac{n}{2}\right)U_{0,y}^2 + \left(m-\frac{n}{2}\right)U_{0,y}V_{0,x} + \frac{1}{2}\left(\lambda+m-\frac{n}{2}\right)V_{0,x}^2 +$$
$$\left(\frac{\lambda^2}{(\lambda+2\mu)} + \frac{8l\mu^2}{(\lambda+2\mu)^2} + \frac{2m(\lambda^2-4\lambda\mu-4\mu^2)}{(\lambda+2\mu)^2} + n\right)U_{0,x}V_{0,y} +$$
$$\left.\left(\frac{\lambda(\lambda+\mu)}{(\lambda+2\mu)} + \frac{4l\mu^2}{(\lambda+2\mu)^2} + \frac{2m\lambda^2}{(\lambda+2\mu)^2}\right)V_{0,y}^2\right] -$$
$$\frac{\lambda\nu z^3}{6(1-\nu)}[\partial_x(\Delta U_0) + \partial_y(\Delta V_0)].$$

Eventually in the 3d order problem we have :

$$0 = \mu U_{3,z} + \mu W_x$$
$$0 = \mu V_{3,z} + \mu W_y$$
$$0 = \lambda(U_{3,x} + V_{3,y}) + (\lambda + 2\mu)W_{3,z}$$

that results in :

$$U_3 = -zW_x$$
$$V_3 = -zW_y$$
$$W_3 = \frac{\nu z^2}{2(1-\nu)}(W_{xx} + W_{yy}).$$

Therefore, the asymptotic expansions (2.70) can be written in the following form:

$$U = U_0(x, y; t) + \varepsilon^2 \frac{\nu z^2}{2(1-\nu)}(U_{0,x} + V_{0,y})_x - \varepsilon^3 zW_x + ...;$$

$$V = V_0(x, y; t) + \varepsilon^2 \frac{\nu z^2}{2(1-\nu)}(U_{0,x} + V_{0,y})_y - \varepsilon^3 zW_y + ...;$$

$$W = -\frac{\nu z}{1-\nu}(U_{0,x} + V_{0,y}) + \varepsilon W(x, y; t) + O(\varepsilon^2)$$

and the problem for U and V may be separated in the leading order from the problem for the displacement W, perpendicular to the plate.

The scale value for time variable can be chosen as

$$T = X\sqrt{\frac{\rho}{E}(1-\nu^2)}.$$

Now, using the relationship (2.71) we derive the problem of the longitudinal deformation waves in the plate, made of Murnaghan's material, in the final form of two coupled equations:

$$
\begin{cases}
u_{tt} - u_{xx} - (1-\nu)u_{yy}/2 - (1+\nu)v_{xy}/2 = \varepsilon\left\{ M_x + N_y + \frac{\nu^2}{12(1-\nu)^2}\left[s_{tt} - \frac{1-\nu}{2}\Delta s\right]_x \right\} \\[2mm]
v_{tt} - v_{yy} - (1-\nu)v_{xx}/2 - (1+\nu)u_{xy}/2 = \varepsilon\left\{ M_y^1 + N_x^1 + \frac{\nu^2}{12(1-\nu)^2}\left[s_{tt} - \frac{1-\nu}{2}\Delta s\right]_x \right\}
\end{cases}
$$
$$(2.72)$$

where $s \equiv U_x + V_y$ is the sum of linear components of strain, Δ is the Laplacian, and the following auxiliary nonlinear differential expressions were introduced as:

$$
\begin{aligned}
M &\equiv \frac{1}{2}\left(u_x^2 + u_y^2 + v_x^2 + v_y^2\right) + su_x + \frac{1-\nu}{2}\left(u_y v_x - v_y^2 - 2u_x v_y\right) + \\
&\quad \beta_1\left[\frac{1}{2}(u_y + v_x)^2 - 2v_y(2u_x + v_y)\right] + \frac{3}{2}\beta_2 s^2, \\
N &= s\left[u_y + \frac{1-\nu}{2}v_x + \beta_1\left(u_y + v_x\right)\right]
\end{aligned}
$$

whilst M_1, N_1 can be found from M, N by means of formal substitution $x \to y$, $y \to x$, $U \to V$, $V \to U$. Two nonlinearity coefficients were introduced also as:

$$
\begin{aligned}
\beta_1 &= \frac{1+\nu}{E}[m(1-2\nu) + n\nu/2] \equiv \beta_1(E,\nu;m,n) \\
\beta_2 &= \frac{2(1-2\nu)(1+\nu)}{3E(1-\nu)^2}\left[(l+2m)(1-2\nu)^2 + 6\nu(1-\nu)m\right] \equiv \beta_2(E,\nu;l,m,n).
\end{aligned}
$$

It can be noted that from the physical viewpoint the 4th order mixed derivatives-terms s_{xtt}, s_{ytt} in the system (2.72)-may be interpreted as the description of the inertia of the transverse displacements of the plate, while the terms, proportional to the 4th order space derivatives from $\Delta s_x \equiv (s_{xx} + s_{yy})_x$ and $\Delta s_y \equiv (s_{xx} + s_{yy})_y$, represent the influence of the deflection of the elementary longitudinal "fibers" due to the deformation caused by the propagation of nonlinear longitudinal strain waves.

Formerly (see, e.g., Nariboli and Sedov, 1970; Potapov and Soldatov, 1984) the analytical study of the two-dimensional nonlinear wave propagation was bounded with the reduction of the system of equations to the Kadomtsev-Petviashvili (KP) equation for $R \equiv \partial U_0/\partial\theta$, $\theta = x - t$:

$$(R_X + c_1 RR_\theta + c_2 R_{\theta\theta\theta})_\theta + R_{YY} = O(\epsilon). \qquad (2.73)$$

It was based on the following assumption that the scales in two (equivalent !) in-plane directions are different: $X = \epsilon x$, $Y = \sqrt{\epsilon}y$, and the following power series expansion is valid:

$$U = U_0 + \epsilon U_1 + O(\epsilon^2); \quad V = \sqrt{\epsilon}[V_0 + \epsilon V_1 + O(\epsilon^2)].$$

After that the KP equation was obtained for the leading order term in R while all terms of order $O(\epsilon)$ were omitted, i.e., the limit $\epsilon \to 0$ was assumed. Obviously, in the limiting case the initial problem becomes singularly perturbed, and the validity of this approximation is not evident.

In Chapter 4 we will find some exact quasi-stationary solutions to our model of the 2D nonlinear waves in the plate governed by the systems (2.66) and (2.72), and valid for periodic boundary conditions on U, V and their derivatives, or, for a localized solution, assume that U, V and their derivatives all tend to zero for $x, y \to \infty$.

2.5   Waves in a wave guide embedded in an external medium

We considered before the wave guides having the lateral surface free of stresses. However, this condition may be realised under special circumstances, and most of wave guides are not free in structures and in nature. The interaction of the nonlinear wave propagating in the wave guide embedded in an external medium is of practical interest and may lead to considerable local stresses and/or energy exchange with the external, active of dissipative, medium, i.e., to dramatic change in the problem statement. We shall discuss the statement of the problem and obtain some basic equations to describe the influence of the energy transfer through the lateral surface of a rod on a nonlinear guided wave. Here most attention will be paid to the mechanisms of wave propagation.

To begin with, we remind the action functional (2.1) and shall consider the second term governing the work of external forces acting on the lateral surface as a reaction to the external medium and leading to the bounded transversal displacements. To calculate the work, we have to define an elementary work on a lateral surface.

First, we consider the guided wave propagation in a prestressed rod embedded into an external solid medium. The simplest and classical model of interaction between the surface and the medium was proposed by Winkler, who considered it as the transversal motion of compressed springs, implemented at the surface, see Figure (2.2):

$$\mathbf{F}(\mathbf{x},t) = -k_1 V. \tag{2.74}$$

It does not cover any interaction between U and V, as it used to in solids, and cannot cover any shear motion. However the so-called sliding contact is widely used in complicated wave problems to model an interface interaction.

Pasternak (1954) considered the interaction with shear motions and derived the improved version of the model:

$$\mathbf{F}_1(\mathbf{x},t) = -k_1 V + k_2 \frac{\partial^2 V}{\partial x^2}, \tag{2.75}$$

where x is the co-ordinate along the rod and k_i are stiffness coefficients of the medium. Therefore the work is modelled as the motion of springs with respect to compression/tension with k_1 and shears with k_2. Schematically it is shown in Figure (2.2), part 1.

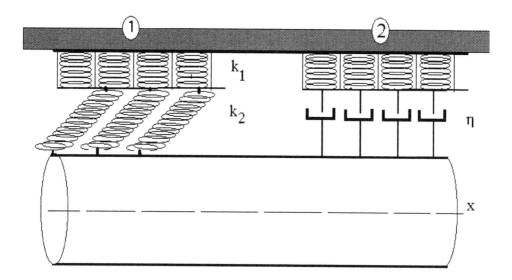

Figure 2.2: Two different models of a 'wave guide - external medium' interaction. The cylindrical rod is embedded in an external medium (gray bar) and interacts according to 1-the Pasternak model; or 2 -the Kerr model. Obviously, the vertical size of an interface is assumed to vanish.

The energy exchange between solids and elastic or viscoelastic foundation was modelled by Kerr (1964). The reaction force is described as a result of an action of elastic and viscous (or active) infinitesimal elements,

$$\mathbf{F}_2(\mathbf{x},t) = -kV - \eta\frac{\partial V}{\partial x}, \qquad (2.76)$$

as it is shown in Figure (2.2), part 2. Here η is the viscocompressibility coefficient, positive for dissipative medium and negative for the active one.

The statement of the problem will be based on the same Murnaghan model as before, and any interaction between a wave guide and an external medium will not result in distorsion of the rod cross section perpendicular to the x-axis; in other words we will use the Love relationship (2.21) and refinements of the plane cross section hypothesis may be made easily. We shall also take into consideration possible prestressing of the rod. The one-dimensional Lagrangian for a hyperelastic rod will be written as:

$$\mathcal{L} = \pi R^2 \left[\frac{\rho}{2}U_t^2 + \frac{\rho}{4}\nu^2 R^2 U_{tx}^2 - \frac{E(1+2U_x^0)}{2}U_x^2 - \frac{\beta}{6}U_x^3 - \frac{\mu\nu^2 R^2}{4}U_{xx}^2 \right],$$

where U_x^0 is the preliminary longitudinal strain component of constant value and other notations are as above. Calculating the elementary work we have:

$$\delta A_1 = \int \mathbf{F}_1(\mathbf{x},t) \cdot \delta \mathbf{U} = (2\pi/3)R^3\nu^2(k_1 U - k_2 U_{xx})_{xx}\delta U,$$

$$\delta A_2 = \int \mathbf{F}_2(\mathbf{x},t) \cdot \delta \mathbf{U} = (2\pi/3) R^3 \nu^2 (kU + \eta U_t)_{xx} \delta U.$$

Then writing the Euler equation, we arrive at the $1 + 1$D equation governing the nonlinear wave propagation in the prestressed elastic rod, embedded in an elastic medium, which lateral surface interacts with it at the interface:

$$U_{tt} - \left[c^2(1 + 2U_x^0) + \frac{2R\nu^2 k_1}{3\rho} \right] U_{xx} = \frac{\beta}{2\rho}(U_x^2)_x + \frac{\nu^2 R^2}{2} \left[U_{xxtt} - \frac{\mu + 4k_2/(3R)}{\rho} U_{xxxx} \right]$$
$$(2.77)$$

When the rod is in contact with the dissipative (active) medium, and the interaction is described by the Kerr model, the equation becomes the DDE with dissipative terms, as follows:

$$U_{tt} - \left[c_0^2(1 + 2U_x^0) + \frac{2R\nu^2 k}{3\rho} \right] U_{xx} = \frac{\beta}{2\rho}(U_x^2)_x + \frac{\nu^2 R^2}{2} \left[U_{xxtt} - c_1^2 U_{xxxx} \right] + \frac{2\nu^2 R\eta}{3\rho} U_{xxt}$$
$$(2.78)$$

We shown that the embedding results in variation of effective values of E and μ, when the rod is inside *the elastic medium*,

$$E_1 \;=\; E(1 + 2U_x^0) + 2R\nu^2 k_1/3,$$
$$\mu_1 \;=\; \mu + \frac{4k_2}{3R},$$

while dealing with the *dissipative medium* together with:

$$E_1 \;=\; E(1 + 2U_x^0) + 2R\nu^2 k_1/3,$$
$$\mu_1 \;=\; \mu$$

we have an additional term $2\nu^2 R\eta u_{xxt}/(3\rho)$ in the right hand side of (2.78) with dissipation constant η, i.e., the external medium can absorb ($\eta > 0$) or transfer ($\eta < 0$) an extra energy to the rod.

Thus in terms of the strain we get:

$$u_{tt} - c_0^2 u_{xx} = \frac{1}{2}\left[\frac{\beta}{\rho}u^2 + \nu^2 R^2(u_{tt} - c_1^2 u_{xx}) \right]_{xx} + 2\eta\nu^2 R u_{xxt}/(3\rho). \qquad (2.79)$$

Dealing with the problem of contact with purely elastic medium one should put $\eta = 0$, whilst the velocities are to be calculated using the modified values of the 2nd order elastic moduli. The influence of an external meduim leads to the increase of both longitudinal and shear wave velocities in comparison with the free rod problem.

This equation was derived under the assumption that the dissipative/active terms are considerable and should be included into the established balance between nonlinear and dispersive terms and the solution was found in (Samsonov, 1988c). Dimensional version will be investigated below in Chapters 4 and 5 for analysis of the influence of interaction in the solitary wave propagation in solid wave guide. If the dissipation is smaller than the two phenomena included, the problem may be solved by means of standard asymptotic expansion on the basis of perturbation of the appropriate DDE solutions.

Introducing the small parameter as above for long wave of longitudinal strain u, assuming a balance between nonlinear, dispersive and dissipative terms (2.79), we obtain the Doubly Dispersion Equation with dissipation, as follows:

$$u_{tt} - u_{xx} = \varepsilon(6u^2 + au_{tt} + bu_{xx} + gu_t)_{xx} + O(\varepsilon^2).\qquad(2.80)$$

where the dimensionless 'viscosity' is defined as $g = \Lambda\eta/(3c_0\rho R)$, while Λ is the characteristic wave length. It was obtained by Samsonov, (1988c); Samsonov and Sokurinskaya, (1989); its analytical features are of considerable interest for the general theory of and solutions to the nonlinear dispersive dissipative p.d.e. that will be discussed in Chapter 3.

Chapter 3

Direct methods and formal solutions

This Chapter is designed to present some mathematical methods of integration of the o.d.e. and to explain intrinsic links between the nonlinear dispersive and dissipative wave propagation problem and the general reduction problem of the classical o.d.e. theory. The focus of this Chapter is mathematical problems, and a reader oriented to nonlinear elastic problems and experiments in solitary waves propagation may omit this chapter, however, in that case, he or she will be asked to accept the mathematical results.

3.1 Nonlinear hyperbolic and evolution equations

The governing equation usually involves a linear (hyperbolic or evolution) wave operator supplied with nonlinear (with respect to an unknown function), dispersive and dissipative terms, all of which may contain variable coefficients, depending on space and time variables. The solution to a nonlinear dispersive dissipative equation constitutes one of the most difficult problems of contemporary mathematical physics. However, many nonlinear p.d.e. can be reduced to an equation with dissipative terms and variable coefficients, into which an independent variable does not enter explicitly, i.e., to the autonomous equation, which gives the ability to obtain a solution in explicit form in some cases.

Recent renewal of interest in possible reductions of nonlinear p.d.e., providing a closed form solution, was motivated, in particular, by the finding of exact particular TW solutions to nonlinear dissipative problems. Some of them were obtained recently for the famous Korteweg-de Vries-Burgers' equation by means of various direct approaches. It revived the interest in application of both classical direct and group theoretical approaches to obtain exact solutions. Results for either the non-dissipative or the non-dispersive problems can be found in papers by Clarkson and Kruskal (1989), Clarkson and Winternitz, (1991), and, also, in the group-invariant solutions in several books. The criterion of equivalence was proposed recently by Arrigo et al. (1993), however comprehensive links between the two methods and the classical theory of nonlinear o.d.e. will have to be established.

We shall find a TW solution instead of any self similar solution due to the following correspondence between self-similar and travelling wave solution of any p.d.e. Usually a self-similar solution to any p.d.e. in $1 + 1D$ written with respect to the unknown function $u(x, t)$ is introduced as an opportunity to represent a solution in a form

$$u(x, t) = U(t) f(X) \tag{3.1}$$

where $X = x/T(t)$ and both U and T are appropriate scaling functions for an unknown and space coordinate.

On the other side, the travelling wave solutions are defined as those depending upon $z = x - Vt + \alpha$ only, namely,

$$u(x, t) = F(x - Vt + \alpha) = F(z). \tag{3.2}$$

Then the obvious substitutions $x = \log z, t = \log(\tau), \alpha = -\log A$, lead to

$$x - Vt + \alpha = \log \frac{z}{A\tau^V} = X,$$

and therefore, to the following transformation

$$u(x, t) = F \left[\log \frac{z}{A\tau^V} \right] = f(X),$$

i.e., any TW solution of the form (3.2) can be expressed as (3.1) and vice versa[1]. We will consider nondissipative problems of generic interest, and after that, turn to dissipative case. As a rule an original wave propagation problem is reduced firstly to the evolution equation, and that way is not trouble free.

A wave propagation problem may be described by means of a nonlinear hyperbolic equation (NHE) or its corresponding reduction to a nonlinear evolution equation (NEE) that will extract the unidirectional wave. At least two main approaches can be used in order to obtain the travelling wave (TW) solution, depending upon the phase variable $z = x \pm Vt$, with velocity V, i.e., the solution of a corresponding o.d.e.:

- direct integration of the NHE, that will be considered below;
- standard reduction to the NEE and integration.

Derived from the Hamilton principle, a nonlinear wave equation, written as

$$u_{tt} - u_{xx} = \epsilon[(N(u) + D(u)]_{xx}. \tag{3.3}$$

where $N(u)$ is a nonlinear term and $D(u)$ is a dispersive one, is quasi hyperbolic: for small ϵ the l.h.s. provides a linear wave equation. For this reason any complete set of boundary and initial conditions for wave problems must be satisfied. In contrast, the corresponding NEE, being a widespread but not always well-grounded reduction of

---

[1] Ya. Zeldovitch in 1956 introduced the self-similarity of the 2nd kind, where $x' = x + a; t' = t + b$ and $a = cb + const$ leads to $u'(x', t') = u(x, t)$.

the NHE, requires a detailed analysis. To obtain the NEE one has to introduce new variables,

$$\begin{cases} \tau = \epsilon^{n+1} t; \\ \xi = \epsilon^n (x - V_0 t), \end{cases} \tag{3.4}$$

and substitute them to (3.3) with given integer n. For example, $N(u) = u^k$ results to the KdV equation for $k = 2$ or to the modified KdV equation for $k = 3$ in the form

$$u_\tau + (u^k)_\xi / k + c u_{\xi\xi\xi} = O(\epsilon), \tag{3.5}$$

or for higher order KdV, when $k > 3$. To solve it one should *omit* the right hand side terms and consider equations without parameters, whereas the initial NHE (3.3) contains small parameter ϵ with highest derivative.

Then one has to conclude that for u and its derivatives being subjected to the boundary conditions:

$$\partial^m u / \partial z^m \to 0, \quad |z| \to \infty, \quad m = 0...3, \tag{3.6}$$

integrations based on NHE and NEE models will provide almost equal results, while if the limiting values of the unknown are different:

$$u \to u_1, z \to +\infty; \quad u \to u_2, z \to -\infty; \text{ and } \partial^{(m+1)} u / \partial z^{(m+1)} \to 0, m = 0...2; |z| \to \infty, \tag{3.7}$$

there will be no equivalence in solutions. The reason of it lies in fact that we deal with the singular perturbation boundary value problem, arising when $\varepsilon \to 0$ in the NEE limit. It will not result in a big difference in a solitary wave solution or bounded periodic (cnoidal) wave solution, while any other solution will differ. In general, only one independent initial condition, corresponding to the originally posed problem, can be taken into account for the NEE, and the question arises for a problem reduced to the NEE, how to satisfy the first independent initial condition $u(t = 0) = u_0(x)$. Obviously, this fact is crucial for the DDE with dissipative terms (see Chapter 2), for which in the limit of KdV-Burgers equation corresponding to the initial NHE with dissipation the conditions (3.7) will not be considered.

That is precisely the case of a singular perturbation problem, usually arising for any equation, containing a highest derivative with a small parameter. The simplest example can be mention here for the o.d.e., governing the static beam deflection problem, see (Coul, 1968; Lomov, 1981). The general statement of linearized problem consists of the perturbed equation:

$$L_\epsilon(u) \equiv \epsilon u_{xxxx} - u_{xx} - p(x) = 0, \tag{3.8}$$

written for a normalized deflection $u(x)$ of a beam under transverse external load $p(x)$, and four boundary conditions for clamped ends, namely, $u(0) = u_x(0) = 0, u(1) = u_x(1) = 0$, whereas the reduced one has the form $L_0(u) = u_{xx} + p(x) = 0$ and only two boundary conditions for function $u(x)$ can be considered.

The exact solution of the problem $L_\epsilon(u) = 0$ can be written formally in the following form:

$$u_\epsilon = u_0(x) + c_1(\epsilon) \exp(-x/\epsilon) + c_2(\epsilon) \exp((1 - x)/\epsilon). \tag{3.9}$$

and consists of the solution $u_0(x)$ of the reduced equation $L_0(u) = 0$ in the limit $\epsilon = 0$ and two boundary layer exponential functions, which have essentially non-zero values only near the boundaries of the interval: $x = 0$ or $x = 1$. The solution, therefore, cannot be expressed as power series with respect to $\epsilon^k, k > 0$ in the vicinity of the singular point $\epsilon = 0$. Hence the exact solution tends to $u_0(x)$ non-uniformly with respect to $\epsilon \to 0$ in the interval $[0, 1]$, and this is exactly the hallmark of a singular perturbation problem arising for any differential equation with a small factor multiplied by the highest derivative of $u(x)$, but not a necessary condition.

In general, if the determination domain $\mathcal{D}(L_0)$ of a limit operator L_0 is greater then the domain $\mathcal{D}(L_\epsilon)$, then the problem is named as the singular perturbed one, and this definition seems to be more rigorous. Indeed, $\mathcal{D}(L_\epsilon)$ consists of any continuously differentiable functions having in $[0, 1]$ four continuous derivatives and satisfying four boundary conditions. The solution $u_0(x)$ of the limit equation $L_0(u) = 0$ does not belong to $\mathcal{D}(L_\epsilon)$: $u_0(x) \notin \mathcal{D}(L_\epsilon)$ for an arbitrary $p(x) \in \mathcal{C}[0, 1]$, furthermore, even for $p(x) \in \mathcal{C}^\infty[0, 1]$ we have $u_0(x) \notin \mathcal{D}(L_\epsilon)$ because of four independent boundary conditions to be satisfied.

The problem becomes even more complicated for an inhomogeneous problem for the NEE with slowly varying coefficients. To solve it one has to use the matched asymptotic expansions, e.g., a double asymptotic expansion (with respect to ϵ and to a parameter describing a smooth inhomogeneity). Therefore it seems to be advantageous to deal with the nonlinear quasihyperbolic equation in physical problems in order to avoid extra difficulties arising in the limit $\epsilon \to 0$.

3.1.1   Travelling wave solutions to the KdV and the DDE equations

Here we shall show how to construct a solution to an o.d.e., which can be useful for a travelling wave solution of the DDE and of the famous KdV equation

$$u_t + 6uu_x + \delta u_{xxx} = 0. \tag{3.10}$$

It is well known that it possesses both solitary wave and periodic cnoidal wave solutions, however in many books they are referered to as the separate sets of solutions. It is easy to demonstrate that the 'general' solution covers both of them. We shall show also that it is definitely discontinuous and may be reduced to any continuous one in a limiting case.

Indeed, the KdV equation results after the phase variable introduction $z = x - Vt$ and integration in a form:

$$c - Vu + 3u^2 + \delta u'' = 0.$$

Multiplying in u' and integrating again we have:

$$(u')^2 = \frac{2}{\delta}\left(-d + cu + V\frac{u^2}{2} - u^3\right), \tag{3.11}$$

that is similar to the Weierstrass equation defining the elliptic complex valued $\wp$ -function:

$$(y')^2 = 4y^3 - g_2 y - g_3 \rightarrow y = \wp(x + x_0; g_2; g_3).$$

To obtain it from (3.11), one may transform u as follows:

$$u = -2\delta y + \frac{V}{6} \tag{3.12}$$

and define the corresponding invariants of the Weierstrass elliptic function as:

$$g_2 = \frac{12c + V^2}{(6\delta)^2}; \quad g_3 = \frac{1}{(6\delta)^3}(108d - 18cV + V^3). \tag{3.13}$$

Therefore, (3.10) has a *discontinuous* general travelling wave solution:

$$u = \frac{V}{6} - 2\delta\wp(z + z_0; g_2; g_3), \tag{3.14}$$

with free parameters V, c, d, z_0, that can be reduced to a *solitary* wave and to a *cnoidal* wave solution by means of appropriate choice of invariants and the $\wp-$ function limits in real axis.

If we seek a solution vanishing at infinity, then we should assume $c = d = 0$ in (3.11) and obtain from (3.13) the particular values of invariants in dependence of velocity V:

$$g_2 = V^2/(6\delta)^2; \quad g_3 = V^3/(6\delta)^3. \tag{3.15}$$

The DDE

$$u_{tt} - u_{xx} = (u^2 + au_{tt} + bu_{xx})_{xx} \tag{3.16}$$

can be reduced to (3.11) also:

$$\frac{(aV^2 + b)}{2}(u')^2 = -\frac{u^3}{3} + \frac{u^2}{2}(V^2 - 1 - c) - du + e, \tag{3.17}$$

where the parameters c, d, e have constant values now. It means that, in general, even travelling wave solutions for KdV equation and for the DDE are discontinuous, depend on the $\wp$-function of the *complex* variable, and additional conditions should be satisfied to obtain a wave solution as a continuous function of the real variable, e.g., as an observable signal from the physical viewpoint.

3.2 Conservation laws

For a nondissipative nonlinear equation one can obtain a set of the so-called conservation laws useful for any testing of numerical simulation techniques and for clear understanding of physics of nonlinear wave propagation in the non-uniform and/or non-homogeneous wave guide.

It is well known that a conservation law for a quantity v expresses a variation of v in a volume W by means of negative flux $f(v)$ across the volume boundary ∂W as

$$\frac{\partial}{\partial t} \int_W v dW = - \int_{\partial W} f(v) d(\partial W), \tag{3.18}$$

whence one can rewrite it in a form of the Gauss theorem $v_t + div f(v) = 0$, or in the differential form

$$D_t + F_x = 0$$

where D is density and F is a flux. For example, starting from the DDE written as

$$4(v_{tt} - v_{xx}) = \varepsilon(6v^2 + av_{tt} - bv_{xx})_{xx}, \qquad (3.19)$$

we introduce D and F by means of the following coupled equations:

$$\begin{cases} D = 4v_t, \\ F = -[4v + \varepsilon(6v^2 + av_{tt} - bv_{xx})]_x \end{cases}$$

that is equvalent to the DDE, when we rewrite it as

$$\begin{cases} 4v_t = -g_x, \\ g_t = -[4v + \varepsilon(6v^2 + av_{tt} - bv_{xx})]_x. \end{cases}$$

Then, using the homogeneous boundary conditions for any solution $v(x,t)$, and its derivatives, vanishing when $\mid x \mid \to \infty$, the integral in (3.18) is a functional on solutions v, and therefore is an integral of the DDE.

Taking into account an invariance of equations for D and F with respect to transformation $t \to -t$ and the boundary conditions, we obtain a set of polynomial conservation laws in the form K_i=const, where

$$K_1 = \int_{-\infty}^{\infty} v dx, \quad K_2 = \int_{-\infty}^{\infty} g dx, \quad K_3 = \int_{-\infty}^{\infty} g dt, \quad K_4 = \int_{-\infty}^{\infty} gv dx,$$

$$K_5 = \int_{-\infty}^{\infty} xv_t dx, \quad K_6 = \int_{-\infty}^{\infty} \left[g^2 + v^2 + \varepsilon(4v^3 + av_t^2 + bv_x^2)/4 \right] dx \qquad (3.20)$$

Some of these quantities provide physical interpretations as the mass conservation (K_1), the momentum conservation (K_5) and the 'energy' conservation (K_6) of a pulse satisfying the above mentioned boundary conditions.

3.2.1   The Hamiltonian structure

To find out whether the DDE is fully integrable by means of the inverse scattering transform (IST), we consider the dimensionless DDE written for a general nonlinear function $F(u)$ as :

$$u_{tt} = [F'(u) + bu_{xx} + au_{tt})]_{xx} . \qquad (3.21)$$

The divergence type system:

$$u_t = v_x,$$

$$v_t = \frac{1}{1 - a\partial^2/\partial x^2} \frac{\partial}{\partial x} [bu_{xx} + F'(u)],$$

as well as the Hamiltonian structure

$$u_t = \frac{1}{1 - a\partial^2/\partial x^2} \frac{\delta H}{\delta v}; \quad v_t = \frac{1}{1 - a\partial^2/\partial x^2} \frac{\delta H}{\delta u} \qquad (3.22)$$

both arise due to the *Hamiltonian operator* introduced as follows:

$$\frac{\partial/\partial x}{1 - a\partial^2/\partial x^2} \qquad (3.23)$$

Conserved polynomial densities can be written now as:

$$K_1 = \int v dx; \quad K_2 = \int u dx; \quad K_3 = P = \int (uv + au_x v_x) dx;$$

$$K_4 = H = \frac{1}{2} \int \left[v^2 + av_x^2 - bu_x^2 + 2F(u) \right] dx,$$

while the last is the Hamiltonian H of the problem, and the variational derivatives are written as:

$$\frac{\delta H}{\delta u} = bu_{xx} + F'(u); \quad \frac{\delta H}{\delta v} = \left(1 - a\frac{\partial^2}{\partial x^2}\right) v = v - av_{xx}$$

We can see that no additional terms to them can make the problem fully integrable by means of the IST method. Moreover, some polynomial densities are not from the set written above:

$$K_5 = \int v dt; \quad K_6 = \int xv_t dx$$

Using a variational derivative definition, (see Newell, 1985):

$$\frac{\delta H_{2n+1}}{\delta q} = \sum_0^\infty \left(-\frac{d}{dx}\right)^n \frac{\partial \tilde{H}(q^{(n)})}{dq^{(n)}}$$

where

$$q^{(n)} = D^n q, \quad D = d/dx; \quad H(q) = \int\limits_{-\infty}^{\infty} \tilde{H}(q, q_x, q_{xx}, ...) dx$$

i.e., for any ε :

$$\lim_{\varepsilon \to 0} \frac{H[q + \varepsilon\eta] - H[q]}{\varepsilon} = \int_{-\infty}^{\infty} \frac{\delta H}{\delta q} \eta dx$$

we conclude that for $n = 0$ the flux corresponds to the trivial translation of an initial pulse:

$$q_x = \frac{\partial}{\partial x} \frac{\delta H_1}{\delta q}; \quad H_1 = \int_{-\infty}^{\infty} \frac{q^2}{2} dx,$$

whereas the flux for the KdV equation is for $n = 3$:

$$q_{t_3} = \frac{\partial}{\partial x} \frac{\delta H_3}{\delta q}; \quad H_3 = \int_{-\infty}^{\infty} \frac{q_x^2 - 2q^3}{8} dx$$

where time-like coordinate t_{2n+1} corresponds to $(2n+1)$-flux, having a linear dispersion relation $\omega(k) = -2(k/2)^{2n+1}$, $n = 0, 1, \ldots$

Note that each functional H_{2n+1}, generating the flux, is the motion integral for another flux, while these integrals commute with each other with respect to the natural Poisson bracket:

$$\{F, G\} = \int_{-\infty}^{\infty} \frac{\delta G}{\delta q} \frac{\partial}{\partial x} \frac{\delta F}{\delta q} dx$$

Evidently, for periodic boundary conditions in flux formuli all integrals are with respect to a period, .Moreover it is valid that :

$$L^n q = \frac{\delta H_{2n+1}}{\delta q}$$

where the nonlocal operator L is:

$$L = -\frac{1}{4} D^2 - q + \frac{1}{2} \int_{-\infty}^{x} dx \, q_x$$

and therefore all fluxes can be written in the Hamiltonian form.

At least, the presense of the mixed derivative does not possess the full integrability of the DDE by means of the IST method.

As a rule, wave solutions of (3.19) differ from those of evolution equations of the KdV-type: $u_\tau + uu_\xi + u_{\xi\xi\xi} = O(\varepsilon)$. Firstly we note the difference between the DDE (3.19) that contains two dispersive terms with opposite signs even in order $O(u^2)$, and the well known Boussinesq equation for incompressible fluid, where due to relation $u_t \cong u_x + O(u^2)$ one term can be expressed through another. Even in a corresponding linearized problem one can reveal the difference in dispersion properties when either u_{ttxx} or u_{xxxx} are taken into account. In Figure 1.2 one can see the dispersion relations, in Figure 1.3 - phase velocities and in Figure 1.4 - group velocities for linear equation, Boussinesq equation, and two versions of DDE respectively.

Note also that the Boussinesq equation of the *1st* kind (containing bu_{xxxx}) possesses the Painlevé property, while it does not allow the separation of variables, whereas the one of the *2nd* kind with au_{xxtt} does not have the Painlevé property, but the variables may be separated.

We should mention the finite non-zero value of phase velocity c_{ph} for $k \infty$ for the DDE, i.e., of the energy transmission velocity. A wave can propagate without energy transfer, when $c_g = \partial\omega/\partial k = 0$, whilst its phase velocity $c_{ph} = \omega/k \neq 0$, and this value of phase velocity can be reached by the DDE wave, at least, formally.

3.3   Some notices in critical points analysis for an o.d.e.

Solutions in terms of the Weierstrass elliptic $\wp$ function, even for problems with dissipation, may also be found. Before to introduce a new approach for finding a solution to a 2nd order nonlinear o.d.e., we shall describe briefly types of singularities for a

1st order o.d.e. It may be based on a multi-valued function dependence of a solution upon initial data.

Let us consider an analytical function $w = f(z)$ of the complex variable z. Its derivative $w' = f'(z)$ may be written as $w' = F(w)$ by means of extraction of z. Of main interest is the case when an equation contains both w and its derivatives in algebraic form with coefficients as analytical functions of z. These equations were studied by Hermite, Poincaré, Painlevé, Fuchs and many others. The 1st order o.d.e. of this kind may be written as

$$P(w(z), w', z) \equiv A_0(w, z)(w')^n + A_1(w, z)(w')^{n-1} + \ldots + A_n(w, z) = 0, \quad (3.24)$$

with P being a polynomial with respect to w and w'.

As a rule only singular points of single-valued functions are considered in the general theory of analytical functions. Unfortunately, integrals of o.d.e. are mostly multi-valued functions. Therefore, following Painlevé, we will introduce some definitions of a singular point being:

critical, where a function changes its value n times, e.g., $z = 0$ for the function $w = \sqrt{z}$ and for $w = \log z$;

non-critical, otherwise, e.g., essentially singular point of a single-valued function or a *pole,* where a function value remains unchanged. The last is the *only* noncritical singular point of algebraic functions of z and is of main interest for our study.

Besides, there are algebraic:

critical singular points, e.g., $z = z_0$, where $w(z)$ is holomorphic in z_0 and may be expanded as

$$w(z) = a + b(z - z_0)^{\frac{1}{n}} + c(z - z_0)^{\frac{2}{n}} + \ldots$$

or a *critical pole of order m* in $z = z_0$, where

$$w(z) = \frac{a}{(z - z_0)^{\frac{m}{n}}} + \frac{b}{(z - z_0)^{\frac{m-1}{n}}} + \frac{c}{(z - z_0)^{\frac{m-2}{n}}} + \ldots$$

Critical points may be non-algebraic. It may be a transcendental singularity, when there is one point of indeterminancy, e.g., $z = 0$ for $w = \log z$, or an essential one, if there is more than one point of indeterminancy, e.g., $z = 0$ for $\exp(1/z)$. Eventually, singular points may be non-isolated, and may constitute singular lines: $|z| = 1$ for

$$w(z) = 1 + \frac{z}{1} + \frac{z^{2!}}{2^2} + , , , + \frac{z^{n!}}{n^2} + \ldots$$

All of these kinds of singularities will not be considered here.

Now we consider singular points of integrals (solutions) of differential equations. The question arises how restrictive is the requirement for (3.24) to be algebraic in the unknown function w and its derivatives. It was Fuchs who proposed an important subdivision of integral's singular points into *movable,* which location depends on initial data, and *unmovable* (fixed). Golubev (1941) found an interesting example of an equation

$$w' = \frac{1}{2wz}$$

having an integral

$$w = \sqrt{\log \frac{z}{C}}$$

for which points $z = 0$ and $z = \infty$ are transcendental singularities, while $z = C$ is the critical algebraic point. Indeed:

$$w = \sqrt{z - C} \left\{ \frac{1}{\sqrt{C}} + \varphi(z) \right\}$$

where $\varphi(z)$ is holomorphic in $z = C$. Then, transcendental points are identical for different values of C, while the algebraic point location varies with C, and therefore, with initial data z_0, w_0, and this point is the movable one. Naturally, some nonlinear o.d.e. have the same feature and the behaviour of these points is of main interest for study of nonlinear o.d.e.

Painlevé in 1887 proved a remarkable theorem:

All integrals of the 1st order o.d.e. (3.24) do not have movable transcendental and essentially singular points, but only algebraic points-poles and algebraic critical points.

Later a useful theorem was proved by Golubev: *Integrals of linear o.d.e. have no critical movable singular points.*

Therefore one can conclude that when a solution in closed form can be found for any particular set of algebraic coefficients in (3.24), it will not have movable essentially singular points, but, probably, poles.

The Fuchs theorem yields the conditions for singularities to be fixed (Fuchs, (1880); cited after Golubev, (1941)):

For (3.24) all critical points are necessarily and sufficiently fixed, if and only if:
A_0 *does not depend on* w : $A_0(w, z) = A_0(z)$;
a polynomial coefficient A_k *with the* $(n - k)$- *degree of the derivative in this equation is at most of degree* $2k$ *in* w.;
the discriminant equation solutions are at the same time the integrals of (3.24);
when the expansion of w' *is*

$$w' = s_0 + b_k (w - w_0)^{k/m} + b_{k+1}(w - w_0)^{(k+1)/m} + \ldots$$

then $k \geq m - 1$.

Integration of an equation with unmovable critical points of genus $p = 0$ can be reduced to an integration of the Riccati equation.

Furthermore, Hermite (1873) proved one of the most important theorem for further analysis:

Any autonomous equation (3.24) $P(w, w') = 0$, *without movable critical points possesses a genus* p *equal either to 0 or to 1, and therefore it has an integral, expressed either in terms of rational functions for* $p = 0$ *or in terms of* $\exp(u)$ *and/or of elliptic functions for* $p = 1$ *(e.g., the Weierstrass function* $\wp$*).*

Eventually we cite an outstanding result by Poincaré (1885):

Any non-autonomous equation (3.24) with unmovable critical points, having the genus $p = 0$, *is reducible to the Riccati equation, while for* $p = 1$ *it is integrable in terms of the Weierstrass* $\wp$ *function after corresponding linear fractional transforms.*

Note that the Riemann surface of genus $p = 0$ is homeomorphic to a sphere, i.e., to a complex plane. Moreover, Hermite proved that the summation theorem is valid *only* for functions, rational with respect to z, to $\exp z$ and for elliptic functions.

Now this is the branch point of our analysis, that may be done in two different ways.

i) Heuristic consideration:

One can use the results by Hermite and Poincaré to introduce *an ansatz* and to find a solution to a 1st or a 2nd order nonlinear o.d.e. as a function rational with respect to $\wp$. It will be based on the statement made by Whittaker and Watson (1927), that *an elliptic function f* is expressed in terms of the function $\wp(z + C)$ and its derivative as

$$f(z) = A(\wp) + B(\wp)\wp', \tag{3.25}$$

with both A and B, being rational functions (or even polynomials) with respect to their arguments. To find A and B the analysis of poles (not all the critical points) was used as a rule. Moreover the solution to an equation is assumed to have only poles.

This approach is *heuristic* and often used to obtain some particular results in dissipative problems of common interest, see, e.g., Samsonov (1988c), Kashcheev (1990), Jeffrey and Xu, (1989), a review by Vlieg-Hulstman and Halford, (1991), Porubov and Velarde (1999), and many others. It may be useful to activate further rigorous studies of the nonlinear 2nd order o.d.e., however we will concentrate here mainly on the formal consideration of the problem.

ii) Rigorous approach:

It consists in analytical study of the problem of critical points analysis and closed form integration for the 2nd order equation:

$$P(w'', w', w, z) = 0, \tag{3.26}$$

that remains under thorough consideration. It is important for physics because one of the main problem in the theory of integration of nonlinear dispersive and dissipative p.d.e. consists in possible integration of corresponding 2nd order nonlinear o.d.e. arisen due to transformation of a p.d.e. to the phase variable dependence. This is just a first step in the theory, and in many problems of nonlinear mathematical physics the 2nd order autonomous nonlinear dissipative o.d.e. :

$$u''_{zz} = a(u)u'_z + b(u) + c(u)(u'_z)^2 + d(u)(u'_z)^3, \tag{3.27}$$

arises as the natural reduction of an initial p.d.e. Group theory analysis will provide a complete set of possible solituions to it, up to date, some results were obtained by Schwarz (1998a,b) and the problem is far from being solved. A similar equation was introduced firstly by Sophus Lie and studied in framework of group theory analysis; for this reason we shall call it as the generalized Lie equation[2].

Here we will show how it can result in closed form solutions in some cases, when the integration of it can be based on a reduction to the 1st order equation.

---

[2] Thanks to Dr. F. Schwarz for the reference to the original Sophus Lie paper.

A successful attempt in the most simple case $c = d = 0$ was made by Ince (1964) who proposed the substitution:

$$u = \mu W(Z) + \nu; \ Z = \varphi(z), \tag{3.28}$$

that allowed to reduce the Lie equation, containing a cubic polynomial $b_3(u)$ with constant coefficients by means of the following option in (3.28)

$$\mu(z) = \exp(\int (a(z)/3)dz), \ \nu(z) = b_2(z)/3, \ Z = \varphi(z) = (-i/\sqrt{2}) \int \mu(z)dz,$$

to the equation

$$W'' = 2W^3 + A(Z)W + B(Z), \tag{3.29}$$

where

$$A(Z) = -\frac{2}{\mu^2}(b_1 - b_2^2/3 - 2a^2/9), \tag{3.30}$$

$$B(Z) = -\frac{2}{\mu^3}[b_0 + b_1b_2/3 + 2b_2^3/27 + (ab_2' - b_2'')/3]. \tag{3.31}$$

Containing the exponential in μ, (3.29) will not have constant coefficients even for constant coefficients in a nonlinear term $b(u)$. It seems to be similar to the 2nd Painlevé equation,

$$u'' = 2u^3 + zu + B$$

if $A(Z) = \alpha Z + \beta$ and $B(Z) - const$, however it cannot be reduced to the second equation otherwise.

3.4 New approach to a solution for an autonomous dissipative nonlinear equation

We consider a problem on finding an exact explicit travelling wave (TW) solution for nonlinear dispersive partial differential equations with dissipative terms. In general, the 4th order p.d.e. includes the nonlinear terms, the dispersive (higher even order terms) and dissipative (odd order terms), that seem to provide a general description of a wave propagation problem for the unknown $u(x, t)$ in space x and time t in a wave guide.

We will find the TW solution that depends only upon the (phase) variable $z = x \pm Vt$ and describes a wave propagation along the x axis in time t with the velocity V.

The interest in the TW solutions for physical applications is based on the simplest approaches to generate and to detect these waves in experiments. Furthermore, TW solutions can be transformed into more general self-similar solution, as it was already shown in Section 3.1.

We will demonstrate how some solutions can be obtained using some new simple reductions and classical transformations. Under certain conditions the autonomous p.d.e. can be reduced to the following second order Lie equation written with respect to an unknown function $u(z)$

$$u''_{zz} = a(u)u'_z + b(u) + c(u)(u'_z)^2 + d(u)(u'_z)^3, \tag{3.32}$$

or even to the simple version of it written as follows

$$u''_{zz} = a(u)u'_z + b(u). \tag{3.33}$$

In the non-dissipative case when $a(u) = 0$, this equation is well known to provide an appropriate integration to obtain a solution in close form, while for dissipative equations (3.32) and (3.33) the problem of finding any exact explicit TW solution is far from being solved.

The truncated Lie equation (3.33), containing both polynomials $b(u) = \sum b_i u^i$ and $a(u) = \sum a_i u^i$, corresponds, after integration, to the standard reductions of travelling wave problem for the classical *KdV-Burgers'* equation for $i = 2$, and constant values of $a(u) = \mu$, α, δ, V,

$$\delta u''' = \mu u'' - \alpha u u' + V u', \tag{3.34}$$

the *modified KdV-Burgers'* equation ($i = 3$, $a(u) = \mu$-const),

$$\delta u''' = \mu u'' + V u' - \alpha u^2 u', \tag{3.35}$$

the *Gardner* equation with dissipation

$$\delta u''' = \mu u'' + V u' - \alpha u u' - \beta u^2 u', \tag{3.36}$$

the equation for nonlinear dissipative waves in solid wave guides (Samsonov, 1982; 1988) or the *double dispersion equation*, where a_0, a and b are given constants,

$$u'' = [u^2 + (aV^2 - b)u'']'' + a_0 u''', \tag{3.37}$$

the *Fisher* equation ($i = 2$, $b_2 = -b_1$),

$$u' = u'' + b_2 u(1 - u), \tag{3.38}$$

and the *Kolmogorov-Petrovsky-Piskounov*, the KPP equation ($i = 3$, $b_2 = -b_1$),

$$u' = u'' + b_2 u(1 - u)^2, \tag{3.39}$$

the *Ginzburg-Landau* equation ($b_2 = 0$),

$$u'' + u' = b_1 u + b_3 u \mid u \mid^2, \tag{3.40}$$

the nonlinear *Schroedinger* equation (NLS):

$$u'' + a_1 i u' + b_3 u \mid u \mid^2 = 0, \tag{3.41}$$

the nonlinear *reaction-diffusion* problem governed by an autonomous o.d.e.

$$D(u)u'' + D_u(u')^2 + V u' + A(u) = 0, \tag{3.42}$$

The usual way to study the nonlinear equations is in qualitative analysis of the behaviour of a system on a phase plane (u', u), however it is not sufficient for finding any exact solution explcitly or testing a numerical one.

Note that the above mentioned Ince substitution (3.28) allows to reduce (3.33) with a cubic polynomial $b(u) = b_3(u)$ and constant coefficients to the equation

$$W'' = 2W^3 + A(Z)W + B(Z), \qquad (3.43)$$

that cannot be reduced to the equation with unmovable critical points.

However, for a polynomial of second degree $b(u) = b_2(u)$ the equation (3.33) can be reduced to either the 1st Painlevé equation for linear $A(Z) : W'' = 6W^2 + Z$, or to the Weierstrass equation: $W'' = 6W^2 + const$, for $\alpha = 0$ in A. The first one is well known to be solved in terms of the 1st Painlevé transcendent, which asymptotic dependence of Z for $\mid Z \mid \to \infty$ is expressed again in terms of the $\wp$ function. The second is solvable in explicit form.

The main question arises how to construct an exact travelling wave solution to dissipative equation (3.33), having, perhaps, some movable critical points, but for nonlinearity of general kind and in explicit form. The Padé approximants technique has been used by Kudryashov (1988) to obtain the partial kink solution for the modified KdV-B equation. Berkovich and Nechaevsky (1983) studied the integrability of the Emden-Fowler equation using a classical approach, that is close to the problem considered. Gundersen (1990) used some similar transformations to obtain TW solutions for the non-dissipative nonlinear equations with constant coefficients, and, particularly, for the sine-Gordon equation.

We propose to find some exact TW solutions by means of reduction of a dissipative nonlinear o.d.e. to the 1st order Abel equation, as it was shown in (Samsonov, 1991; 1993, 1995). Some of these solutions for higher order o.d.e. and coupled equations can be obtained in explicit form.

3.5   A general theorem of reduction

Let us consider the problem of finding some exact solutions of eqs. (3.27) and (3.33). The theorems by Hermite (1873) and by Painlevé (1887) mentioned above lead to the conclusion that when a solution in closed form to (3.27) can be found for any particular set of algebraic coefficients, it will not have movable essentially singular points, but, probably, poles.

The problem of exact explicit solution of a nonlinear dissipative wave equation can be transformed now into the following question: is it possible to extract the equations, which can be integrated in terms of the $\wp$ -function and may have movable singularities like poles?

For reduction the equation (3.27) to a low order equation we propose the following:
Theorem.
The second order autonomous ordinary differential equation, the Lie equation (3.27),

$$u''_{zz} = a(u)u'_z + b(u) + c(u)(u'_z)^2 + d(u)(u'_z)^3,$$

containing nonlinear and dissipative terms with *arbitrary* functions $a(u)$, $b(u)$, $c(u)$, $d(u)$, can be reduced to the 1st order Abel equation:

$$v'_u + b(u)v^3 + a(u)v^2 + c(u)v + d(u) = 0. \tag{3.44}$$

or in the normal form:

$$y'_\xi = y^3 + I(\xi). \tag{3.45}$$

Proof

is straightforward:

introducing the new unknown function v via differential substitution $v(u) = 1/u'_z$, we have

$$u''_{zz} = -\frac{v'_u u'_z}{v^2} = -\frac{v'_u}{v^3}$$

and reduce (3.27) to the lower order equation with respect to v, that yields the non-autonomous Abel equation of the 1st kind (3.44) after a little algebra[3]. For continuous $c(u)$ and continuously differentiable $a(u)$ and $b(u)$ the following substitution

$$w(u) = \exp(\int(-c + a^2/(3b))du, \quad \xi(u) = -\int b(u)w^2(u)du, \quad v = w(u)y(\xi) - a/(3b),$$

is well known to reduce equation (3.44) to the normal form (3.45) with function $I(\xi)$, defined by an expression

$$w^3 b(u)I(\xi) = d - ac/(3b) + 2a^3/(27b^2) - (a/3b)'_u. \tag{3.46}$$

Therefore we reduced the integration of the initial 2nd order problem to the integration of the 1st order Abel equation (3.44). The theory of the Abel equation and abelian integrals is not complete and the equation is not integrable in closed form in general case.

Its closed form solution may be based on either using a particular type of function $I(\xi)$ or the substitution for $d(u) = 0$, i.e., for (3.27) that does not contain the highest order dissipative term, as it was found by Lemke (1920). Consequently, the equation (3.44) is transformed into the non-autonomous 2nd order o.d.e.

$$t^2 \xi''_{tt} + g(\xi) = 0, \tag{3.47}$$

where different transformations were used

$$w = \exp(-\int c(u)du), \quad \xi = \int w(u)a(u)du, \quad v(u) = w(u)y(\xi), \quad g(\xi) = -w(u)b(u)/a(u),$$

together with the differential substitution

---

[3]Formerly this theorem was proven (Samsonov, 1991) for a polynomial $b(u)$ only and for $c = d = 0$; moreover, a different transformation was used.

$$\xi'_t = 1/(ty(\xi)). \tag{3.48}$$

We note at this step that the initial substitutions that would have led to the normal form $y'_\xi = y^3 + I(\xi)$ seems to be rather useless.

Using this approach, the equation of 2nd order (3.47) arises again, but fortunately it does not contain now the first derivative of the unknown function (i.e., the *dissipative* term), as it was in equation (3.27). Any explicit solution of equation (3.47) provides a solution of the initial problem, however the closed form solution appearance is limited by whether the function $y(\xi)$ can be defined explicitly from (3.48) for any t.

For (3.33), in particular, we reduce (3.44) to the following equation

$$v'_u + [b(u)v + a(u)]v^2 = 0. \tag{3.49}$$

Remembering the Fuchs theory, one can see that its main result is not valid for the Abel equation obtained because $k = 1$ in (3.44), while the highest degree is 3, therefore we proved also

Corollary 1

The Lie equation of general kind does not have fixed singularities only.

It means that the heuristic approach mentioned above should be supplied with the critical points analysis.

Moreover, we have

Corollary 2

Only for $p = 0$ or 1 both w and $w' = W(z)$ are the single valued functions of the Abel integral of the kind $z = \int dw/W$, or generally, $z = \int R(W, w)dw$. If $p > 1$, both w and W have no singular points, but singular lines or other manifolds. For the p.d.e. it means even the strongest dependence on initial and boundary conditions.

The theory of abelian integrals is connected with the problem on uniform curves, and for $p = 1$ it is well known that the simplest curve of genus $p = 1$ is the third order curve defined with an equation $s^2 = 4t^3 - g_2t - g_3$, i.e., with the equation formally defining the Weierstrass elliptic function.

Corollary 3

The Riccati equation does not follow from any dissipative nonlinear problem, while the Abel equation does, as was shown above.

3.6 Dissipative equations with polynomial nonlinearity

Many TW propagation problems of mathematical physics are governed by a short version of the nonlinear Lie equation (3.33) that allow the employment of further reduction. Applying the approach described, we can obtain the second order equation (3.47), whereas in this case

$$v(u) = v(\xi); \; \xi = \int a(u)du; \; g(\xi) = -b(u)/a(u),$$

and the last term does not necessary contain an exponential term. Therefore if $g(\xi)$ has no singularities and $\xi(t)$ is expected to have poles only, one can obtain a solution to the equation in terms of elliptic functions.

Further drastic simplification for finding an exact solution is based on an assumption that $g(\xi)$ is reducible to a polynomial, as it follows from many applications. In the general case this leads the equation (3.47) to the *dissipative* 2nd order equation with *constant* coefficients. Malmquist (1914), see also Golubev (1941), for arbitrary *polynomials a, b, c, d* proved that if the equation (3.44) is not of the Riccati type, then any of its integrals is a rational function.

Therefore if both $a(u)$ and $b(u)$ are polynomials $a(u) = a_k(u)$ and $b(u) = b_m(u)$ of any order with respect to the function u, then (3.33) leads to the generalised Emden-Fowler equation by means of substitution (3.48), namely, to the form

$$t^2 \xi'' + g_{m-k}(\xi) = 0, \tag{3.50}$$

i.e., the non-autonomous non-dissipative equation with rational nonlinearity $g(\xi)$ of $(m - k)$ degree.

To demonstrate a usage of the approach let us assume that $g(\xi)$ is a polynomial of order n : $g(\xi) = \sum \gamma_i \xi^i$, with coefficients defined by a constant dissipation coefficient $a(u) = a$ and a polynomial $b(u) = \sum b_i u^i$. For $\xi = \xi(t)$ we apply the following substitution:

$$\xi = Ct^p F(y) + D, \; y = t^q. \tag{3.51}$$

with arbitrary constants C and D. Our object is to obtain the autonomous equation from (3.50). The substitution of ξ and $g(\xi)$ leads to the autonomous nonlinear equation with *constant* coefficient for *arbitrary n*

$$F'' - KF^n = 0, \tag{3.52}$$

if and only if the following additional conditions are valid :
for autonomity:

$$p = 2/(n + 3); \tag{3.53}$$

for absence of the 1st derivative of F:

$$q = 1 - 2p = (n - 1)/(n + 3), \; n > 1; \tag{3.54}$$

and for an appropriate constant coefficient with the highest degree of $F(y)$

$$K \equiv C^{n-1}(n + 3)^2 \gamma_n / (n - 1)^2, \tag{3.55}$$

while to exclude the lower degrees of F one should satisfy the set of additional conditions for polynomial coefficients:

$$\gamma_0 + \gamma_1 D + D^2(\gamma_2 + \gamma_3 D + \ldots + \gamma_n D^{n-2}) = 0, \tag{3.56}$$

$$\gamma_1 + D(2\gamma_2 + 3\gamma_3 D + \ldots + n\gamma_n D^{n-2}) = p(p - 1) = -2(n + 1)/(n + 3)^2, \tag{3.57}$$

$$\gamma_2 + D(3\gamma_3 + 6\gamma_4 D + \ldots + \gamma_n n(n-1)D^{n-2}/2!) = 0, \tag{3.58}$$

etc.; these equalities follow from the corresponding requirements for each binomial coefficient in $g(\xi) = g[Ct^p F(y) + D]$ to be equal to zero. Otherwise these conditions extract those nonlinear dissipative equations that are integrable explicitly by means of this approach. The solution found to the dissipative nonlinear o.d.e. is not necessarily unique.

3.6.1   Square and cubic polynomial nonlinearities

The most often considered nonlinear equations having polynmial nonlinearity of order $n = 2$ and $n = 3$ are of special interest for the theory of nonlinear waves in dissipative systems. They can be integrated now in closed form.

For $n = 2$ we have $p = 2/5$, $q = 1/5$, and from (3.54) one can define D from the equation

$$\gamma_1 + 2\gamma_2 D = p(p-1) = -6/25, \tag{3.59}$$

and the additional relationship for γ_i is to be valid:

$$\gamma_0 + \gamma_1 D + D^2 \gamma_2 = 0, \tag{3.60}$$

The constant C is used to make an appropriate value for coefficient for F^2. Assuming $C = 6/(25\gamma_2)$ the equation (3.52) becomes the Weierstrass equation $F_{yy}'' - 6F^2 = 0$, that has a *general* solution in terms of the Weierstrass elliptic function, namely, $F = \wp[y + y_0; 0; g_3]$ with invariants $g_2 = 0$ and g_3; this type is usually called the equianharmonic limit for the function $\wp$.

From (3.56)-(3.58) we get the following conditions for coefficients γ_i:
i) if $\gamma_0 \neq 0$ then $D = [p(p-1) - \gamma_1]/(2\gamma_2)$ and it results to the corresponding relationship for $\gamma_0, \gamma_1, \gamma_2$

$$4\gamma_0\gamma_2 = [\gamma_1 + p(p-1)][\gamma_1 - p(p-1)] = \gamma_1^2 - 36/625, \tag{3.61}$$

that should be valid to obtain the travelling wave solution to the initial problem.
ii) if $\gamma_0 = 0$, then either $D = 0$, and therefore it should be $\gamma_1 = p(p-1) = -6/25$, or $D = -\gamma_1/\gamma_2$ and therefore $\gamma_1 = -p(p-1) = 6/25$.

Similarly for $n = 3$ we have $p = 1/3$, $q = 1/3$, and (3.55), (3.56-3.58) yield $\gamma_1 = -2/9$, $C^2 = 2/(9\gamma_3)$, then the result leads to the equation

$$F_{yy}'' - 2F^3 = 0, \tag{3.62}$$

evidently integrable in terms of elliptic function $\wp$ if the following relationships for γ_i are valid:
iii) if $\gamma_0 \neq 0$ then $D = -\gamma_2/\gamma_3$ and $\gamma_0\gamma_3 = \gamma_2\gamma_1$, provided that $\gamma_1 = -2/9 - \gamma_2^2/\gamma_3$.
iv) if $\gamma_0 = 0$ then either $D = 0$, and therefore $\gamma_1 = p(p-1) = -2/9$, $\gamma_2 = 0$, or $\gamma_1 = 0$, $2\gamma_3 = -9\gamma_2^2$, and $D = -\gamma_2/\gamma_3$. Both cases provide the travelling wave solutions to the Gardner equation (3.36), as it will be shown below.

The important question arises how to perform an inverse transformation in (3.48) and to write both functions $y(\xi)$ and, respectively, $u(z)$ in explicit form. Comparing the substitutions, we find

$$\xi'_t = 1/(tv(\xi)) = au'_t$$

and

$$v(\xi) = v(au) = 1/u'_z,$$

therefore for arbitrary function u the following equation must be satisfied:

$$atu'_t = u'_z, \tag{3.63}$$

from which the independent variable t is defined as $t = \exp(az)$, and finding ξ from (3.47) one can perform the inverse transforms to obtain an explicit TW solution in closed form.

Finally, we obtain for any problem governed by the nonlinear dissipative equation (3.33) with a constant a, and nonlinearity $b(u) = \gamma_0 + \gamma_1 u + \gamma_2 u^2$, (here $\gamma_i = b_i/a^{i+1}$) the solution under additional condition $4\gamma_0\gamma_2 = \gamma_1^2 - p^2(p-1)^2$, $p = 2/5$, is

$$u = D/a + (C/a)y^2 \wp(\beta y + y_0),$$

and in explicit form,

$$u = -(6/25 + \gamma_1)/(2a\gamma_2) + (6/(25a\gamma_2)) \exp(2az/5)\wp[\exp(az/5) + z_0; 0; g_3], \tag{3.64}$$

that contains two arbitrary parameters, z_0 and g_3 to satisfy two boundary conditions. Recently this solution was obtained for the dissipative double dispersion equation (3.37) in (Samsonov, 1995). In particular, if the value $g_3 = 0$ is given also by a boundary condition, this solution results in the exact algebraic jump (kink) TW solution

$$u = A + B\exp(2az)(\exp az + z_0)^{-2}, \tag{3.65}$$

where both factors A, B, are defined as in (3.64). The kink wave as a partial solution was obtained in (Samsonov, 1988c) by means of the Painlevé equation analysis and in many papers (see Jeffrey and Xu, 1989; Kudryashov, 1988; McIntosh, 1990) some of those were reviewed by Vlieg-Hulstman and Halford (1991).

We remark that, following the Ince book (1964), one can reduce the equation $u'' = au' + b_2 u^2 + b_1 u + b_0$ to the Weierstrass form under appropriate conditions for coefficients, that provides the kink solution (3.65) free from movable critical points. Here the general solution (3.64) with poles was found to the o.d.e. considered, containing two arbitrary parameters to satisfy the boundary conditions. Then one can study how (3.64) tends to (3.65), following the variation of parameters, when $g_3 \to 0$.

Let us assume $D = 0$ for simplicity. The general solution is discontinuous, and discontinuities are condensed at inifinity. For values of g_3 far from zero (Figure 3.1) the first discontinuity of u is close to $y = 0$ (i.e., for $z \to \infty$).

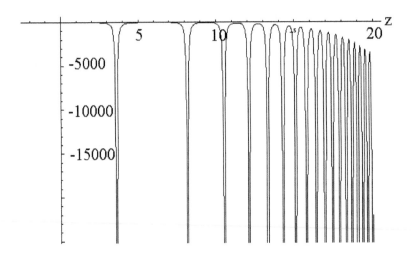

Figure 3.1: General discontinuous solution to the Lie equation in terms of $\wp$ function for $g_3 = 1$.

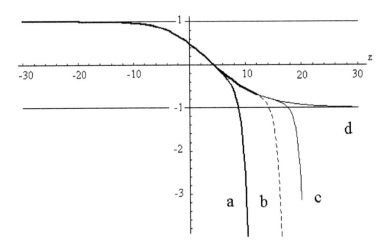

Figure 3.2: Discontinuous solution tends to the smooth kink non-uniformly with respect to g_3: a) $g_3 = 10^{-5}$; b) $g_3 = 10^{-7}$; c) $g_3 = 10^{-9}$; d) $g_3 = 0.0$.

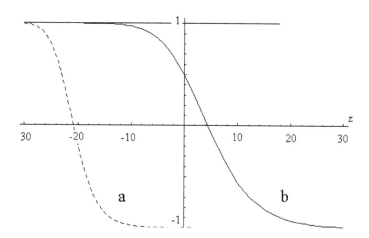

Figure 3.3: Kink location depends on the value of the initial shift z_0 : a) $z_0 = 10^{-4}$; b) $z_0 = 1.0$.

Further calculating, we see (Figures 3.2a, b) that the less is g_3, the longer is the distance between the origin and the first singularity, while for large y the solution exhibits an infinite set of discontinuities, condensing when z tends to infinity. Even for $g_3 = 10^{-9}$ (see Figure 3.2c) the function $u(y)$ tends to an almost constant value for values of y near the origin. When $g_3 = 0$ (see Figure 3.2d) all discontinuities tend to infinity and disappear, whereas in the vicinity of $z = 0$ the smooth kink solution (3.65) arises, if the figure scale is appropriate to detect it.

The initial shift z_0 defines an initial position of a centre of the kink, see Figure 3.3. Therefore, the behaviour of $u(y)$ is singular, indeed, while the difference between two sets of coefficients of the initial problem, that provide either a singular solution or a smooth kink solution is surprisingly small. This observation might be instructive also for any numerical simulation of a solution.

Considering both smooth and discontinuous solutions of nonlinear dissipative problems, one can study the dynamics of development of discontinuity. From the physical viewpoint a solution, having discontinuity far enough from the origin, can be of interest itself because many nonlinear dissipative equations are good approximations of corresponding asymptotic expansions with respect to a small parameter ϵ, given by physics, that are valid for time and space values less then $1/\epsilon$.

Similarly, we can obtain the exact solution for (3.33) with cubic nonlinearity $b(u) = \gamma_0 + \gamma_1 u + \gamma_2 u^2 + \gamma_3 u^3$ and constant dissipation coefficient a. Assuming $\gamma_0 = 0$, integration of (3.62) yields the rational fractional function written in terms of $y = \exp(az/3)$ as

$$u = -\gamma_2/3\gamma_3 + (2\gamma_3)^{-1/2} y\wp_y'/\wp(y, g_2, 0)], \qquad (3.66)$$

while additional conditions $iii)$ or $iv)$ for coefficients are to be satisfied. The solution (3.66) contains the $\wp$-function in its pseudolemniscate limit, i.e., $\wp(y, 1, 0)$, which can be obtained after corresponding transformation. In particular, for $\gamma_2 \neq 0$ it constitutes the exact solution for the Gardner equation (3.36), while for $\gamma_0 = \gamma_2 =$

0, $9\gamma_1 = -2a^2$, $\gamma(u) = \gamma_1 u + \gamma_3 u^3$, a kink-shaped solution arises, as it was found in (Samsonov, 1991).

In addition, we may formulate:

Corollary 4

The general Lie equation with cubic polynomial nonlinearity and dissipative function, both having constant coefficients, cannot be reduced to the equation with unmovable critical points;

and can say that:

The general nonlinear dispersive dissipative equation possesses a travelling wave solution in explicit form, if both nonlinearity and dissipation functions are polynomials, and some additional conditions for coefficients and the wave velocity are valid.

3.6.2 Physical interpretation of additional conditions for the travelling wave existence

We have shown that both equations for $n = 2$ and $n = 3$ have the closed form solutions in terms of the Weierstrass $\wp$ - function. Now we will discuss briefly the physical meaning of additional restrictions for coefficients in order to integrate the nonlinear dissipative equation explicitly.

Finding the TW solution to the Korteweg-de Vries-Burgers equation (3.34) $u_t + \alpha u u_x + \delta u_{xxx} = \mu u_{xx}$ or to the double dispersion equation (3.37), we introduce the phase variable $z \equiv x - Vt$ and obtain respectively the Abel equation $v'_u + b(u)v^3 + a(u)v^2 = 0$ with $a(u) \equiv a = -\mu/\delta$, $b(u) = b_0 + b_1 u + b_2 u^2$, where for (3.34) $b_1 = V/\delta$, $b_2 = -\alpha/(2\delta)$ and b_0 is a constant of integration. Typical boundary conditions when $\mid z \mid \to \infty$, under which the TW solution is to be found are: $u \to u_i$, whilst u', $u'' \to 0$. It leads to the relationship:

$$(u_1 - u_2)[b_1 + b_2(u_1 + u_2)] = 0, \qquad (3.67)$$

then either values of u are equal at infinity or they should satisfy the equation $u_1 + u_2 = 2V/\alpha$.

Finally we write the TW solution (3.64) in terms of initial parameters of physical problem as

$$u = (\mu^2/(\alpha\delta))\left[6/25 + V\delta/\mu^2 - (12/25)\exp(2\mu z/(5\delta))\wp[\exp(\mu z/(5\delta)) + z_0; 0; g_3]\right],$$
$$(3.68)$$

while the additional relationship follows from (3.61) in the form $(V\delta/\mu^2 - 6/25)(V\delta/\mu^2 + 6/25) = 2\gamma_0\alpha\delta^2/\mu^3$ and represents the balance between dissipation ('viscosity') and dispersion, and wave velocity as the necessary condition for existence of the TW solution.

Assume that the boundary conditions in a physical problem are such that $u_1 = u_2 \equiv u_0$. Then $b_0 = -u_0(b_1 + b_2 u_0)$ and $b_0 = 0$ if $u_0 = 0$, therefore (3.68) does not contain a constant, and the balance is given by $25V\delta = -6\mu^2$.

Vice versa, if $u_1 + u_2 = -b_1/b_2$, we have $b_0 = u_1 u_2 b_2 = -\alpha u_1 u_2/(2\delta)$, and at the same time b_0 is defined by (3.61), then $4ab_2^2 u_1 u_2 = b_1^2 - p^2(p-1)^2$, or in physical variables

$$(V/\mu)^2 - 36/625 = -\alpha^2 \delta^3 u_1 u_2 / \mu^5, \tag{3.69}$$

that can be equal to zero if any of boundary values u_i vanish at infinity. Therefore, the wave front $u(z)$ propagates with prescribed velocity and switches over an initial statement of the physical system into another one.

3.7 Elliptic function solutions to higher order problems

The reduction to the 2nd order o.d.e. and even to the Abel equation is not possible in many problems of considerable physical interest, that are described, e.g., by means of the 3d order and the higher order equations containing source-like terms. One can mention as appropriate examples both the FitzHugh-Nagumo problem for nerve pulse transmission and the water wave problem with surface tension being taken into account.

Nevertheless, the results obtained above allow us to put forward a heuristic idea for these problems to find some exact explicit solutions of nonlinear o.d.e (3.27) in the form (4.49) of an elliptic function:

$$f(z) = A(\wp) + B(\wp)\wp',$$

The procedure seems to be quite simple. One should substitute any of the expressions assumed to be appropriate as A and B, from (4.49) into a *differential* equation under study, make independently the coefficients at each order of $\wp$ and $\wp'$ equal to zero, and obtain the set of coupled *algebraic* equations for them. When a solution of the system exists, perhaps, under certain restrictions for coefficients, we obtain a solution of initial nonlinear p.d.e. We mention the book by Kashcheev (1990), containing some preliminary results obtained by a heuristic approach usage. Moreover, any mathematical symbolic software can be used effectively for performing a large amount of tedious calculations at this stage.

Let us formulate, following Engelbrecht (1991), the FitzHugh-Nagumo model, describing a problem of nonlinear nerve pulse transmission. We consider the coupled nonlinear equations for the unknown function $u(x,t) \in [0,1]$, that are presented nowadays in the Hastings form as

$$\begin{cases} u_t = u_{xx} + F_3(u) - R, \\ \quad R_t = \epsilon(u - bR), \\ F_3(u) = u(1-u)(u-m), \end{cases} \tag{3.70}$$

where R is the recovery function, $1/2 > m > 0$, $0 < b < 1/M$, $M \equiv \max(dF/du)$ and $F_3(u)$ is the cubic source term[4]. Looking for any explicit TW solution, we reduce these

---

[4] The *2nd* order equation $u'' = u' + u(1-u)^2$ represents the KPP equation (or at least, the Fisher equation with cubic nonlinearity) and often misnamed the FitzHugh-Nagumo equation.

equations to the *third* order o.d.e., corresponding to the FitzHugh-Nagumo model, as follows:

$$u''' + a_1 u'' + u'(a_2 + a_3 u - 3u^2) + a_4 u + a_5 u^2 + a_6 u^3 = 0. \tag{3.71}$$

where $a_1 = V - \epsilon b/V$, $a_2 = -m - \epsilon b$, $a_3 = 2(1+m)$, $a_4 = -\epsilon(1-mb)/V$, $a_5 = -\epsilon b(1+m)/V$, $a_6 = \epsilon b/V$. To find an explicit solution we assume $u = A(\wp) + B(\wp)\wp'$, $\wp = \wp(z + z_0)$. Substituting it to (3.71) and separating terms into two parts proportional to $\wp$ and to $\wp'$, respectively, yields

$$B'''Q^2 + 6B''RQ + B'(3R^2 + 48\wp Q) + 12B(Q + \wp R) + a_1(A''Q + A'R) + a_2(BR + B'Q)$$

$$+a_3(AB'Q + ABR + A'BQ) - 3(2ABA'Q + A^2 B'Q + B^2 B'Q^2 + A^2 BR + B^3 QR)$$

$$+a_4 A + a_5(A^2 + B^2 Q) + a_6(A^3 + 3AB^2 Q) = 0, \tag{3.72}$$

and

$$A'''Q + 3A''R + 12A'\wp + a_1(B''Q + 3B'R + 12B\wp) + a_2 A' + a_3(AA' + BB'Q + B^2 R)$$

$$-3(A^2 A' + A'B^2 Q + 2A'BB'Q + 2AB^2 R) + a_4 B + 2a_5 AB + a_6(3A^2 B + B^3 Q) = 0. \tag{3.73}$$

Here primes denote differentiation of A and B with respect to their arguments $\wp$, and the well known relationships were used to express the derivatives of $\wp$ in terms of $\wp$: $Q(\wp) \equiv (\wp')^2 = 4\wp^3 - g_2\wp - g_3$; $R(\wp) \equiv \wp'' = 6\wp^2 - g_2/2$, $\wp''' = 12\wp'\wp$, where g_2, g_3 are invariants of $\wp$. To explain the algorithm itself we assume for simplicity that both functions A and B are polynomials with respect to $\wp$, $A = A_k(\wp)$, $B = B_n(\wp)$, then each term in equations will be a polynomial also. Further one should estimate the highest degree of $\wp$ in each equation in dependence upon the intervals in which are k and n. If two or more terms of these degrees are in the equation and these values of $k = k_1$ and $n = n_1$ are the same for both equations, then the necessary balance is achieved between two different terms at least. Simple calculations lead to the following estimations for those terms in (3.72), that contain either A or B only:

$$k_1 = 3k, \text{for } k \geq 1, \text{and } k_1 = k + 1, \text{for } k \leq 0;$$

$$n_1 = 3n + 5, \text{for } n \geq -1, \text{and } n_1 = n + 3, \text{for } n \leq -1;$$

as well as for (3.73)

$$k_1 = 3k - 1, \text{for } k \geq 0, \text{and } k_1 = k, \text{for } k < 0;$$

$$n_1 = 3n + 3, \text{for } n \geq -1, \text{and } n_1 = n + 1, \text{for } n < 1;$$

Further, for these particular values of k and n from the fixed interval one should estimate the highest degree term containing both A and B. When both equations (3.72) and (3.73) can be satisfied by the same values of k and n, the degrees of polynomials are defined. For the problem of TW solution to the FitzHugh-Nagumo model we have found the following pairs of values of k and n:

$$\left\{ \begin{array}{l} k = 0 \\ n = -1 \end{array} \right. ; \left\{ \begin{array}{l} k = 1 \\ n = 0 \end{array} \right. ; \left\{ \begin{array}{l} k = -1 \\ n = -1 \end{array} \right. ; \left\{ \begin{array}{l} k = -2 \\ n = -1 \end{array} \right. ,$$

then several partial TW solutions of the problem can be found in one of the following simplest forms:

$$u = A_0 + B_0 \wp'/(\wp + C); \quad u = A_0 + B_0\wp + C_0\wp'; \tag{3.74}$$
$$u = (A_0 + B_0\wp)^{-1} + \wp'/(\wp + C_0); \quad u = (A_0 + B_0\wp)^{-2} + \wp'/(\wp + C_0); \tag{3.75}$$

respectively. To find any of them explicitly one should solve the corresponding coupled algebraic equations, defining the constants A_0, B_0, C_0, wave parameters and coefficients of recovery function R and source term F. This set arises after collecting coefficients of monomials with respect to each degree of $\wp$ in equations resulted from (3.72), (3.73).

Recently the exact solution to the FitzHugh-Nagumo equation was found in (Samsonov, 1993) having the form $u = A_0 + B_0\wp'/(\wp + C)$ with both constants and source term subjected to some special conditions.

The approach can be used for the *fourth* order (a non-dissipative example for simplicity) o.d.e. of the form

$$u'''' = u'' + u^k - u. \tag{3.76}$$

Following the proposition (4.49) with $B = 0$ and polynomial A($\wp$) of order m, i.e., $u = A_m(\wp(z))$, we obtain the condition $m(k - 1) = 2$, that for cubic nonlinearity $k = 3$ yields $m = 1$, hence the exact solution is

$$u = A_0 + A_1\wp(\gamma z), \tag{3.77}$$

while for $k = 2$ the equality $m = 2$ is valid, and therefore

$$u = A_0 + A_1\wp(\gamma z) + A_2\wp^2(\gamma z). \tag{3.78}$$

The last formula provides the ability to obtain TW solution to the equation, derived by Hunter and Scheurle (1988) for refined description of surface tension influence on water waves, where numerical solution was obtained, having an oscillation structure. Additional relationships for free values of invariants of $\wp$ and coefficients of an equation provide the construction of the solution.

We have shown that many dissipative nonlinear TW propagation problems, reducible to the generalized Lie equation, result in the 1st order Abel equation, consequently. In some nonlinear quasi *hyperbolic* problems of physical interest it allows to obtain the *elliptic* function solutions.

It seems likely that the approach under consideration is based on the fact that any autonomous equation $v'(z) = q(v)$ with a polynomial $q(v)$ can be solved in terms of $\exp(\alpha z)$ and of the Weierstrass function $\wp(z + c)$ among all special functions of mathematical physics, because any derivative of both functions can be expressed as polynomials with respect to the function itself. For this reason the above mentioned reduction to the Abel equation should be noted as a suitable example.

3.8 Example of a nonlinear reaction-diffusion problem

Various problems of nonlinear wave formation in population dynamics, kinetics of thin film growth, nanostructure technology, combustion, flame propagation and heat transfer in nonlinear medium, see, e.g., Nicolis and Prigogine (1977); Murray (1989); Kurdyumov et al. (1981), can be reduced to a single nonlinear reaction-diffusion (NRD) equation. In general, such an equation governs the natural system evolution under competitive 'income' and 'outcome' processes. The problem of exact solutions has been of considerable interest for a long time, renewed recently due to difficulties in numerical simulation, e.g., of nonlinear thin film growth, of wire-like nanostructure formation in kinetics of spontaneous islanding on a substrate, see Dubrovsky and Kozachek, (1995); Kurdyumov et al. (1981). Another interesting example of direct and explicit solutions provides a nonlinear reaction-diffusion problem

$$u_\tau = (D(u)u_x)_x + A(u). \qquad (3.79)$$

containing arbitrary diffusion $D(u)$ and absorbtion $A(u)$ functions. Here we will show how to calculate various closed form solutions; for details see Samsonov (1998); Samsonov and Gursky (1999).

Introducing a phase variable $z = x \pm V\tau$, where $V \geq 0$ is a constant wave velocity, we obtain a nonlinear o.d.e. of second order:

$$D(u)u'' + D_u(u')^2 + Wu' + A(u) = 0, \qquad (3.80)$$

where $W = \mp V$, and primes here denote the differentiation with respect to z. The equation is similar to (3.27) and in general it may be studied as above. Of special interest is the stationary case . The *stationary* solutions can be easily obtained in *implicit* form even for both arbitrary reaction and diffusion functions. Indeed, when $u_\tau = 0$ is assumed in a general p.d.e. and in (3.27) $a(u) = 0$, a new substitution $w(u) = (u')^2$, $' \equiv d/dx$, leads to the Bernoulli equation and to the linear 1st order equation of the form:

$$w'(u) + 2\left[\log D(u)\right]' w(u) + 2A(u)/D(u) = 0, \qquad (3.81)$$

having the general solution for *arbitrary* diffusion and adsorbtion functions, and:

$$w(u) = -\frac{2}{D(u)}\int A(u)D(u)du + \frac{C_1}{D^2(u)}, \quad C_1 - const. \qquad (3.82)$$

The inverse transform leads to the *exact implicit general* solution of the problem under consideration

$$x = C_2 + \int \frac{D(u)du}{\sqrt{C_1 - 2\int A(y)D(y)dy}}, \quad C_2 - const. \qquad (3.83)$$

However, this abelian (hyperelliptic) integral cannot be calculated explicitly[5], and the problem remains how to invert all the transformations and to obtain an explicit

---

[5] e.g., if $C_1 \neq 0$ and/or $A(u)D(u)$ is a polynomial of order more than three.

solution in terms of $u(x,t)$. Note that the solution (3.83) can be studied and inverted numerically in a simple way.

Again, the polynomials $A(u)$ and $D(u)$ allow an explicit integration of (3.42) in stationary case. Moreover, the integral can be inverted, in general, in terms of the Weierstrass elliptic function $y = R[\wp(x)]$, where R is a rational function that provides an explicit solution to the problem, see also Section 20.6 in the book by Whittaker and Watson, (1927; 1962). According to the Riemann surfaces theory (Springer, 1957) even for a genus $g = 2$ an inverse function $w(\cdot)$ always exists and has no singularities at an appropriate Riemann surface S except a pole of order 2. In particular, an *explicit* solution to (3.42) with polynomials A and D can be found in terms of the Weierstrass elliptic function $\wp$ as $y = A_1 + B_1\wp(x + x_0, g_2, g_3)$, where $x_0, g_2, g_3, \ A_1, B_1-$const. Writing the diffusion and absorbtion functions as polynomials:

$$D(y) = 1 - ky, \ A(y) = -y(a - ly + cy^2). \tag{3.84}$$

and using an approach proposed above to the stationary version of (3.42)

$$(D(y)y')' + A(y) = 0, \tag{3.85}$$

where the coefficients are given by (3.84), we find easily an *explicit closed* form solution

$$y(x) = A_1 + B_1\wp(x + x_0, g_2, g_3) = \frac{5kl - 3c}{12ck} - \frac{10k}{c}\wp(x + x_0, g, h) \tag{3.86}$$

where x_0 is an arbitrary constant, and the second and the third invariants of $\wp$ are defined by coefficients of (3.84) as follows:

$$g_2 \equiv g = \frac{-3c^2 + 2ckl + 5l^2k^2 - 16ack^2}{240k^4}, \tag{3.87}$$

$$g_3 \equiv h = \frac{27c^3 - 27lc^2k - 15cl^2k^2 + 72ac^2k^2 - 25l^3k^3 + 120alck^3}{43200k^6}; \tag{3.88}$$

obviously $k \neq 0$ for a non-constant diffusion in (3.84). It should be noted that the solution (3.86) was obtained under minimal restrictions for coefficients of the initial physical problem.

3.8.1 Discontinuous solutions

The solution (3.86) is, in general, discontinuous and semibounded from below or above. Some classical limits of $\wp(\cdot)$, in which it can be easily calculated may arise under additional conditions for coefficients. An interesting case occurs when $l = 0$, that leads (3.86) to following form :

$$y(x) = -\frac{1}{4k} - \frac{10k}{c}\wp(x + x_0, g_2, g_3)$$

with invariants $g_2 = -c(3c + 16ak^2)/(240k^4)$, $g_3 = c^2(3c + 8ak^2)/(4800k^6)$. In particular, for $c = -16ak^2/3$ one has $g_2 = 0$, $g_3 = -32a^3/675 \neq 0$ and $B_1 = +15/(8ak^2)$, that provides an equianharmonic case for the Weierstrass function $\wp$. For another possible value of c equal to : $c = -8ak^2/3$, one can obtain $g_2 = 4a^2/45$, $g_3 = 0$ and $B_1 = +15/(4ak^2)$, that defines the lemniscate limit of $\wp$. Moreover, when physical parameters of the problem under consideration allows to get $5lk - 3c = 0$, that results in $A_1 = 0$, we obtain from (3.86),(3.87), (3.88):

$$B_1 = -6/l; \ g_2 = -al/(9k), \ g_3 = al^2/(108k^2)$$

If, *in addition*, we have $a = 0$ in (3.84), the solution (3.86) tends to the limit:

$$y(x) = -\frac{6}{l}\wp(x, 0, 0) = -\frac{6}{lx^2}.$$

The discontinuous solutions can be useful for tentative validation checks during numerical simulation.

3.8.2   Bounded periodical and solitary wave solutions

Of main physical interest are the bounded continuous solutions to the NRD problem, and we obtain them as appropriate limits of the solution (3.86) after some non-trivial algebra.

Aiming to exact and explicit solutions, we introduce 2ω and $2\omega'$ as primitive periods of $\wp$, defined by the condition for an imaginary part: $Im(\omega'/\omega) > 0$, the discriminant $\Delta = g_2^3 - 27g_3^2$, and the roots $e_i = \wp(\omega_i)$, $i = 1, 2, 3$ of a characteristic equation $4e^3 - g_2e - g_3 = 0$, where $\omega_1 = \omega$, $\omega_2 = \omega + \omega'$, $\omega_3 = \omega'$. Following the reduction of the doubly periodic function $\wp$ to a set of single periodic Jacobi elliptic functions, one can write the following relationship between $\wp$ and cn, sn with modulus M as:

$$sn^2(z\sqrt{e_1 - e_3}; M) = \frac{e_1 - e_3}{\wp(z) - e_3}; \ cn^2(z\sqrt{e_1 - e_3}; M) = \frac{\wp(z) - e_1}{\wp(z) - e_3}; \ M \equiv \frac{e_2 - e_3}{e_1 - e_3};$$
$$(3.89)$$

Moreover, both the summation theorem for $\wp(z + \omega_i)$ and the shift rule

$$\wp(z + \omega_i) = e_i + \frac{(e_i - e_p)(e_i - e_j)}{\wp(z) - e_i}; \ \wp[x + (2n + 1)\omega'] = \wp(x + \omega'); \ . \quad (3.90)$$

where $i, j, p = 1, 2, 3$, n is integer, and $i \neq j \neq p$, should be used in calculations. It is known the behaviour of $\wp$ depends on sign and value of Δ, that allows to extract two cases of main interest.

3.8.3   Autosoliton solution

Let us consider the value of a discriminant $\Delta = 0$, then one of the periods is infinite: $\omega = \infty$ or $\omega' = i\infty$ (the trivial case $\omega = -i\omega' = \infty$ will be excluded). The first case $\omega = \infty$ corresponds to $e_1 = e_2 \neq e_3$, and introducing $e_1 = e_2 \equiv E$, we have

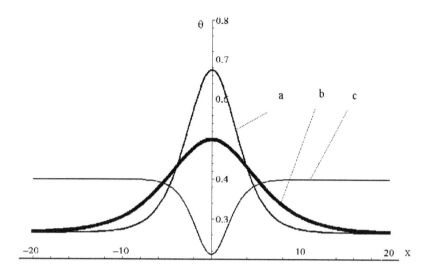

Figure 3.4: Autosoliton solutions.

$e_3 = -2E$, $g_2 = 12E^2$, $g_3 = -8E^3$, $\omega' = i\pi/\sqrt{12E}$, therefore for any real constant x_0 we can take $x_0 + \omega'$ and have from (3.86,3.89):

$$
\begin{aligned}
y(x) &= -\frac{1}{4k} + \frac{5l}{12c} - \frac{10k}{c}\wp\left(x + x_0 + \omega'; \ g_2, \ g_3\right) = \\
&= -\frac{1}{4k} + \frac{5l}{12c} - e_2\frac{10k}{c} - (e_3 - e_2)\frac{10k}{c}cn^2\left(\sqrt{e_1 - e_3}\,(x + x_0)\mid M\right) = \\
&= -\frac{1}{4k} + \frac{5l}{12c} - \frac{10k}{c}E + \frac{30k}{c}E\,cn^2\left(\sqrt{3E}\,(x + x_0)\mid M\right). \quad (3.91)
\end{aligned}
$$

Due to $M = 1$ in this case and the limiting value $cn(u, 1) = \cosh^{-1}(u)$, we obtain *the exact autosoliton solution* $y(x)$ to the original problem (3.85) for given A and D:

$$
y(x) = -\frac{1}{4k} + \frac{5l}{12c} - h\frac{10k}{c} + \frac{30kh}{c}\cosh^{-2}\left(\sqrt{3h}\,(x + c_0)\right), \quad (3.92)
$$

Therefore the variation of $y(x)$ due to reaction and diffusion from a constant value represents the bounded localized structure. In Figure 3.4 the negative autosoliton solution is shown for particular physical parameters of a solitary wave $y(x) \equiv 0.531 + A_1 + B_1\cosh^{-2}(\sqrt{3h}x)$:

$$
A_1 = 10^{-17} \propto 0; \ B_1 = -0.3128; \ \sqrt{3h} = 0.2418;
$$

The case $\omega' = i\infty$ corresponds to $e_1 \neq e_2 = e_3$ and does not lead to a bounded solution.

3.8.4 Periodic bounded solutions in case $\triangle > 0$

In this case there is a pair of primitive periods 2ω, $2\omega'$ such that ω is real and ω' is pure imaginary semiperiods of $\wp$ function. Then $\wp(z)$ will have real values only on

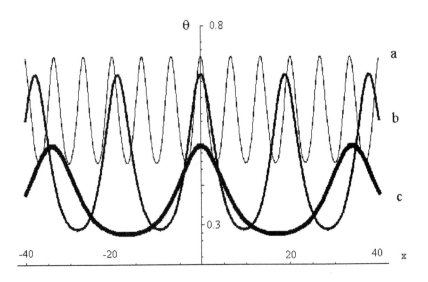

Figure 3.5: Periodic solutions to nonlinear reaction-diffusion equation.

the following lines of a complex z-plane: $Re(z) = 2s\omega$, $iIm(z) = 2p\omega'$, where s, p are integers, which lines agree with the period grid, as well as on the lines of semiperiods: $Re(z) = (2s+1)\omega$, $iIm(z) = (2p+1)\omega'$. However, $\wp(z)$ is discontinuous on real axis and on each line obtained by its shift along the period grid described above. The real bounded periodic solution $y(x)$ can be found only if the real axis x will be shifted to the semiperiods line, and the shift constant $N = c_0 + (2p+1)\omega'$ with any real c_0 and integer p. Furthermore, due to (3.90) for any x and p one has $\wp(x + (2p+1)\omega') = \wp(x + \omega')$, and taking $N = c_0 + \omega'$ without loss of generality, where c_0 is real and ω' is purely imaginary, it provides a transition of the real axis x to the line $i\,Im(x + N) = \omega'$ of the complex variable $x + N$, and the function $\wp$ will have an argument on this line now.

Assuming $\triangle > 0$, all roots e_i, $i = 1, 2, 3$ are real and different, $e_1 > e_2 > e_3$, $e_1 > 0$, $e_3 < 0$ and from the periodic discontinuous solution (3.86) we obtain the bounded *cnoidal wave solution* for polynomials $A(y)$ and $D(y)$ from (3.84) in the form:

$$
\begin{aligned}
y(x) &= -\frac{1}{4k} + \frac{5l}{12c} + \frac{10k}{c}\wp(x + c_0 + \omega_3; g_2, g_3) = \\
&= -\frac{1}{4k} + \frac{5l}{12c} - e_2\frac{10k}{c} - (e_3 - e_2)\frac{10k}{c}cn^2\left(\sqrt{e_1 - e_3}(x + c_0)\mid M\right).
\end{aligned}
\tag{3.93}
$$

An example of this wave is shown in Figure 3.5 for the following parameters of the cnoidal wave solution: $y(x) \equiv A_1 + B_1 cn^2(x \mid M)$:

$$A_1 = 0.22; \quad B_1 = 0.22; \quad \sqrt{e_1 - e_3} = 0.1553; \quad M = 0.9130$$

No bounded real solutions $y(x)$ were found from (3.86) when $\triangle < 0$.

Chapter 4

Nonlinear strain waves in elastic wave guides

We have derived from Chapter 2 the equations governing the longitudinal waves propagation in various wave guides and studied them formally in Chapter 3. Now we shall consider main conclusions and restrictions resulted from theoretical analysis, and, eventually, estimate explicitly all that is necessary to construct long strain solitons in solids and describe real physical experiments in succcessful generation and observation of strain solitons in solids. As a result of theoretical investigations we shall present the description of the first successful experiments in bulk soliton observation in solids.

4.1 Features of longitudinal waves in a rod

We shall start with analysis of the initial pulse parameters necessary for the solitary wave generation in the one-dimensional problem of waves in a nonlinearly elastic rod and analysis of the experimental setup requirements.

From the very beginning of the research on the nonlinear dynamics of large waves in solids, it became quite clear that the nonlinear guided wave propagation problem is complex, very important for physics and engineering and requires thorough mathematical analysis.

As it was shown in Chapter 2 the complete description of a 3-D nonlinear wave in continuum remains difficult; that is why the initial 3-D problems are usually reduced to the 1-D form in order to get the simplest but qualitatively new *analytical* solutions. Very often the linearization of a problem was done, however, it turns out to be unsatisfactory from the genuine physical point of view, because the ratio of a finite deformation and its linear part is determined by the displacement gradient and its variation in time, see, e.g., Lurie (1980), Pleus and Sayir (1983), McNiven and McCoy (1974), Engelbrecht (1983), Shield (1983).

Probably, Nariboli and Sedov (1970) derived firstly both the dimensional KdV and KdV-Burgers equations for nonlinear wave in cylindrical rod using heuristic explanation and re-discovered the solitary wave solution for KdV equation in nonlinear dynamics of solid wave guide. Mechanical stress-strain relation analysis was used lately by Wright (1984) who applied main ideas of nonlinear elasticity to wave problems in elastic wave guides and granular materials. Ostrovsky and Sutin (1977) obtained independently the KdV equation and the soliton solutions, using the Hamilton principle for the (truncated) Lagrangian density. Engelbrecht (1981) studied the nonlinear dynamics of elastic waves thoroughly, using the internal energy expansion in power series with respect to strain tensor invariants and the KdV equation analysis.

However, we shall show that the KdV approximation is too restrictive for theoretical estimations and unsatisfactory for experiments. The solitary wave solution to the dimensional DDE (2.54) for the longitudinal strain component $u \equiv U_x$

$$u_{tt} - c_0^2 u_{xx} = \frac{1}{2} \left[c_2 u^2 + \nu R^2 \left((\nu - 1)u_{tt} + c_0^2 u_{xx} \right) \right]_{xx} \tag{4.1}$$

where $c_0^2 = E/\rho, \quad c_2 = \beta/\rho, \quad \beta = \beta(E, \nu; l, m, n)$ is found to be:

$$u = A \cosh^{-2} \frac{1}{\Lambda} \left(x \pm t \sqrt{c_0^2 + \frac{A\beta}{3\rho}} \right). \tag{4.2}$$

where the values of both squares of velocity[1] V and of the length must be positive:

$$V^2 = c_0^2 + \frac{A\beta}{3\rho}, \quad \Lambda^2 = 2(\nu R)^2 \left(\frac{(\nu - 1)}{\nu} + \frac{3E}{|A\beta|} \right). \tag{4.3}$$

Then the velocity of the compression wave should belong to the following interval :

$$c_0^2 < V^2 < \frac{c_0^2}{1 - \nu}, \tag{4.4}$$

for conventional solids, having the Poisson coefficient inside the interval : $1/2 \geq \nu > 0$, or to two semi-infinite intervals

$$V^2 < \frac{c_0^2}{1 - \nu}, \text{ or } V^2 > c_0^2 \tag{4.5}$$

for 'expanding' solids with $\nu < 0$.

Some details about construction and behaviour of such solids are in the well known papers by Lakes (1987, 1992). Honeycombs with inverted hexagonal cells, synthetic composite laminates and microporous polymers do have negative Poisson's ratio and may be described as structural elements.

For most metals with $\nu \propto 0.3$ the velocity of compressive soliton lies in the interval $(1.0; 1.2)$ of values of the so called rod's velocity c_0, at the borders of the interval either the soliton amplitude tends to 0 or its length. Outside the interval lies the 'dead zone' of velocities, in which solitons cannot propagate at all. The KdV approximation of the DDE leads to *incorrect* estimations for V and Λ :

$$c_0 < V < \infty, \quad \Lambda = R\sqrt{E/(A\beta)}, \tag{4.6}$$

---

[1]We study the u−wave here only and the displacemet V will not be considered in order to avoid confusion.

the main disadvantages are that the velocity of solitons in conventional solids would not be restricted from above and the pulse length value would be estimated incorrectly. Correct values of wave length and velocity are of crucial importance for successful experiments in soliton generation in solids. The type of the solitary wave in dependence on sgn β, ν and relative velocity V^2/c_0^2 is presented in the table below for various materials.

Wave type	β	ν	V^2/c_0^2
Compression	< 0	> 0	$(1;\ 1/(1-\nu))$
Extension	> 0	> 0	$(1;\ 1/(1-\nu))$
Extension	> 0	< 0	$(1;\ \infty)$
Compression	> 0	< 0	$(0;\ 1/(1-\nu))$
Compression	< 0	< 0	$(1;\ \infty))$
Extension	< 0	< 0	$(0,\ 1/(1-\nu))$

In particular, we can see that for $\nu > 0$ there is no subsonic ($V < c_0$) solitary wave of compression in any elastic material, but only the transonic soliton: $V > c_0$ exists. In addition, we know that for $V \approx c_0$ the solitary wave amplitude vanishes, hence it may be one of the reasons why there is no reference to the successful experiments in ultrasonic generation of long strain solitary wave in elastic wave guide. Therefore our first choice can be in the conventional ($\nu > 0$) material subjected to the compressive transonic initial pulse.

We begun with the experiments in soliton observation in solids in 1985 (Samsonov and Sokurinskaya, 1985; Dreiden et al. 1986). At that time there was only one report concerning the successful experiments in *envelope* solitary waves in solids, namely, in the envelope flexural solitary waves detected in a thin cylindrical open-ended shell by Wu et al. (1987). The shell was excited by an acoustical beam impact loading a lateral surface generated by horn driver. Elastic waves occured were flexural having both axial and circumferential components. On a brass-made coaxial ring the capacitive transducers were mounted to measure the flexural radial displacement amplitude of the shell. Flexural wave pulses were made by 5 wave train pulses at a frequency of 1120 Hz. Three oscilloscope were used to record the response of the shell arosen in the envelope solitons arriving at 120 degrees and 240 degrees and travelling clockwise at the velocity of 26 m/s. These two packets overlap at 185 degrees, and a temporal shell response at 120 degrees shows the eventual dispersion of the pulse. The problem of standing envelope soliton propagation was claimed to be governed by the nonlinear Schroedinger equation for the flexural displacement W as

$$aW - bW''_- + 2cW|W|^2 = 0$$

having the hyperbolic secant solution

$$W = \gamma_1 \cosh^{-1}\left[\gamma_2(x - Vt)/R\right]$$

where all parameters a, b, c, γ_i, V depend upon geometry and elasticity of a wave guide. The stability of the envelope wave against nonlinear collision and reasonable agreement between theory and experimental data allowed to interprete these waves as envelope solitons of flexural displacement.

The density long solitary waves remained unobservable.

The main problem in the soliton dynamics in solids is in the tentative solution of non-stationary problem of the generation of a solitary wave from an initial *transonic* pulse. We estimated the parameters of the *elastic* soliton, however we are unable to solve the generation problem analytically, and for this reason we shall investigate this process numerically and in experiments.

Dealing with elasticity, we have to avoid any plastic flow of deforming material and to look for reversible finite deformations that results in the following restriction for the finite strain tensor component:

$$\left|\sqrt{1 + 2C_{xx}} - 1\right| < e_0$$

where e_0 is a yield point. In terms of strain u it means that:

$$\left|u + \nu^2 R^2 u_x^2/2 + O(\varepsilon)\right| < e_0$$

and the proper candidate to generate a solitary wave seems to be an initial weak shock, the deformation in which will not be greater than e_0.

Next questions arising are how to generate weak shock and how to calculate the wave amplitude? The amplitude A of a soliton should be chosen in order to balance nonlinearity and dispersive features of the wave guide that leads to the following estimation of the solitary wave parameters via the elasticity of the wave guide:

$$|A| = \frac{\rho\nu^2 R^2}{\Lambda^2}\left|\frac{V^2 - c_1^2}{\beta}\right|\left|\frac{f''}{f^2}\right|$$

where f is the variable part of the soliton solution $u = Af(\theta)$ in (4.2) and the value of f''/f^2 is of order unity and depends on the initial condition, e.g., for the compressive strain soliton it is equal to 6, see Samsonov and Sokurinskaya (1988) for further details. Again one may underline the difference between models based on the KdV and the DDE approaches: in the last model both dispersive terms could compensate each other when the velocity value is close to c_1, which breaks down the balance necessary for any soliton existence. This possibility is not recognizable in the KdV approach.

Eventually our estimations lead to the following intervals for parameters of tentative transonic solitary compressive wave in the nonlinearly elastic rod:

$$V > c_0; \quad |A| < e_0; \quad \beta < 0; \quad \Lambda > \nu R \sqrt{\frac{(\nu - 1)}{\nu} + \frac{3E}{e_0 |\beta|}}.$$

where Λ is the characteristic pulse width at the level of $A/2$. We have to choose now an elastic material with appropriate ratio of nonlinearity E/β.

Table 4.1. Elastic moduli of 2nd and 3d order.

Moduli, $(*10^{-11} N/m^2)$	λ	μ	ν_1	ν_2	ν_3
Material					
Brass L62; fired grain 2 μm	0.77	0.47	-6.3	-5.4	-5.0
Brass LS59; fired grain 2.5 μm	0.89	0.38	0.1	-1.5	-4.9
Duralumina D16 grain 2 μm	0.7	0.28	-7.1	-6.4	-4.2
Duralumina D16 grain 8 μm	0.57	0.29	-4.8	-3.5	-3.3
Duralumina D16 grain 12 μm	0.58	0.28	-0.4	-4.8	-5.1
Bronze BrOF	1.07	0,44	-1.4	-4.3	-1.7
Bronze- Be	1.05	0.49	-4.0	-1.7	-0.6
PMMA	0.39	0.19	-0.08	-0.07	0.05
Quartz; melted	0.16	0.28	0.54±0.13	0.93±0.1	-0.11±0.03
Pyrex glass	0.13	0.28	2.6±**3.7**	-1.68±**1.38**	1.05±**0.9**

Moduli, $(*10^{-11} N/m^2)$	λ	μ	l	m	n
Material					
W; doubly backed	1.63	1.37	-5.25	-4.72	-4.29
W; single backed	0.75	0.73	-2.67±0.17	-2.50±0.30	-2.15±0.30
Polystyrene SD-3	0.018	0.01	-0.19±0.03	-0.13±0.03	-0.10±0.02

The 3d order elastic moduli data are widely used in industry but rarely published, moreover only a few data are available for the 4th order moduli. Static experimental arrangements are inappropriate for the high order elastic moduli measurements aimed to nonlinear wave experiments.

The values of moduli given in the Table above were cited after the publication by Savin et al. (1976), and after a handbook by Frantsevich et al. (1982). Note the large intervals of errors for some values of (ν_1, ν_2, ν_3), that is hardly recognised properly even now, e.g., in (Erofeev, 1999), however it may considerably change the estimations of wave parameters propagating in any particular material, e.g., as it is in Pyrex glass.

Values of nonlinear elastic moduli are measured mostly in dynamic (ultrasonic) experiments, and for alloys strictly depend on size of grains, as a rule, the smaller the grains the more are the absolute values of the 3d order moduli. It seems to be logical-a material consisting of large grains would be roughly described by a continuum approximation.

For some materials only the Murnaghan moduli data are available. The relationships between the 3d order moduli introduced by Lame (ν_i), by Landau (A, B, C) and by Murhaghan (l, m, n) are given below:

$$l = \nu_2 + \nu_1/2; \ m = \nu_2 + 2\nu_3; \ n = 4\nu_3$$

$$A = n; \ B = m - n/2; \ C = l - m + n/2$$

The following relationships for the 2nd order elastic moduli are useful in applications:

$$E = \frac{\mu(3\lambda + 2\mu)}{\lambda + \mu}, \quad \nu = \frac{\lambda}{2(\lambda + \mu)}$$

From (4.3) it is clear that the ratio E/β should be sufficiently small (i.e., the nonlinearity sufficiently large) to generate a soliton in a rod. It is given in Table 4.2. for several materials:

Table 4.2. Types of observable strain solitons.

Material	E/β	Type of nonlinear wave proposed
Brass L62; fired	**-0.165**	compression
Steel Rex 53	-0.14	compression
Duralumina D16	-0.09	compression
Brass LS59-1	-1.09	compression
PMMA	+0.3	tensile waves
Quartz; melted	+0.16	tensile waves
CuNi 9 alloy	-0.15	compression
Polystyrene	**-0.06**	compression

Evidently, brass, steel and duralumina are good examples of material for the compression soliton generation experiments among opaque solids, while polystyrene is the most appropriate one among optically transparent media.

4.2   Experiments in nonlinear waves in solids

To plan the real physical experiments in nonlinear wave propagation in solids we have to solve various problems besides the theory developed above. The mathematical study mentioned in the Introduction is partially fulfilled, and now we will have to determine further steps to construst the solitons in solids, in particular, we have to describe:

- type of the generation setup;

- an approach to generate an initial pulse with flat front;

- types of interaction of a pulse and a surface of a wave guide;

- an approach to measure both amplitude and velocity in solids;

- types of observation setup.

Each point is in fact a branch point, that means that some decisions will be criticial for successful experiments. We shall report below the results obtained in 15 years of work of our research group in the Ioffe Institute of the Russian Academy of Sciences in generation, observation and detection of bulk solitons in solids.

It is well known that the optical methods are preferable to study transparent optical phase inhomogeneities, e.g., the bulk density waves we are looking for. They allow not only to visualize inhomogeneity but also to determine its parameters. On the other side, being contactless, they do not introduce any disturbances in condensed matter under study, see Dreiden et al. (1994). All the optical methods record the changes of refractive index inside an object, when studying optically transparent phase inhomogeneities.

Shadowgraphy is more convenient to record a considerable refractive index gradient (i.e., the wave front image), for example, caused by a shock wave propagation. It was shown theoretically in our case that a strain soliton is a propagating long bulk density wave of small amplitude. Interferometry is the most appropriate for such waves' study because it allows for observation of even small refractive index variations and measure them with sufficient accuracy.

Holographic interferometry will be used in our experiments. It has several advantages in comparison with the conventional optical interferometry. In particular, limitations to the optics quality are considerably lower because wave fronts to be compared pass through the same optical path. For this reason both waves are distorted *to the same extent* and possible defects in optical elements and experimental cell will be subtracted in the holographic image and do not affect the resulting interference pattern.

However the choice of an optical recording method allow to study, in general, only those elastic materials that are transparent for the given light wave length. The approach must be modified for an opaque material investigations.

Before performing the soliton experiments we shall study the formation of an initial pulse and its features. A setup should include the generation arrangement, the observation apparatus and the registration and measurement unit.

We dealt with the generation and the interaction of shocks with a plane interface of liquid and solids. The weak shock formation under conditions of optical breakdown at the sample surface was studied by Amiranoff et al. (1985), and by means of thermooptical excitation caused by powerful laser pulse (Karabutov et al. (1984)). Both experiments followed by fracture of the sample, i.e., the deformation became inelastic, and it is inappropriate for our purposes.

The shock generation without any damage of the sample was proposed and studied by Dreiden et al. (see papers in 1986 and further publications), and it will be considered below. The setup shown in Figure 4.1 consisted of pulse generation unit (1, 5, 6, 10, 12), optical interferometry setup (2, 3, 4, 9, 11, 13, 14) for observation and registration using both schlieren shadowgraphy (4, 13, 14, 7, 8, 9) and holography (4, 11, 9 and 4, 7, 8, 9). Observation and registration units were incorporated in one apparatus.

For schlieren photography registration of experiments the optical lag line was used, consisted of mirrors (13) and (14). Schlieren doubly exposed photography is sensitive to the gradient of refraction index and used as a rule for registration of sharp waves, e.g., the shocks, while optical holography, being more precise, can demonstrate smooth and prolonged waves as the fringe shifts, and in this setup one can use both methods simultaneously.

In shadowgraphy the laser radiation is expanded by an optical system and divided by the semi-transparent mirror in two parts. The first is going through the shock wave directly, whilst the second is going trough the optical lag line and through the shock after it, providing in our setup the time lag $\tau = 133$ nanosec. The shadowgram of shock is recorded in the film as the sequence of light rings on dark background. The doubly exposed schlieren photograph is shown in Figure 4.2, where one can see two images of the shock shifted at the distance, that was passed by the right running wave during the time interval between two consequent pulses. The distance between two rings allows for the calculation of the velocity of the wave. The optical system was focused sharply on the central line of the entire cut-off of the sample, and for this reason the (transparent) cut-off plane perpendicular to the plane of picture is out of focus and looks like two vertical light lines.

The initial shock in the liquid is generated due to the optical breakdown in the point indicated as a small light bubble and has a velocity more than the sound velocity in liquid. The maximal size of the bubble is ca. 0.2 mm.

One can see that due to interaction with the interface the wave in solids accelerates because of the increase of a wave resistance (or the acoustical resistance $z = \rho c$) in solids. The initial wave is spherical, both reflected wave and wave in solid are distorted. When z in solids is much greater than in liquid, (for melted quartz used in our experiments $z = 133.0 \times 10^5$ $kg/(m^2 s)$) there is no wave in solids, and almost all the energy reflected from the interface, see Figure 4.3 .

To check that the excited strain wave possesses indeed the main features of the soliton, like the shape conservation, it is necessary to follow in observations its propagation along an extended elastic wave guide. However, the more absorbing is a wave guide material for linear elastic waves, the much shorter distance is to be sufficient to detect the constant shape wave propagation, if it will exist. Polystyrene absorbs well both linear and shock elastic waves and it is widely used as an acoustic power pulse absorber, see, e.g., Shutilov (1980); Bushman et al. (1996).

Using our experimental arrangement we may deal with optically transparent sufficiently nonlinear solids, having the wave resistance close to the surrounding liquid. We may choose now the water cell and the rod made of the polystyrene SD-3 as an appropriate wave guide to generate and observe strain solitons. These media are transparent

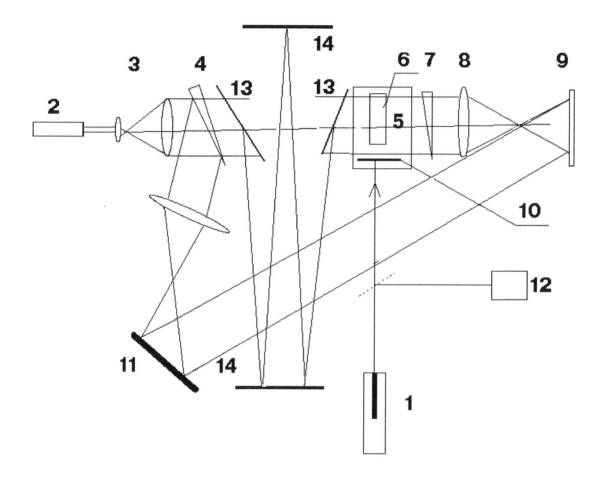

Figure 4.1: Optical scheme of experimental arrangement for weak shocks and strain soliton generation in solids. It contains two pulse ruby lasers (1,2), two lens systems (3,8), semi-transparent mirrors (13) for shadowgraphy and optical wedges (4, 7) for holography, the register unit (9), where photo or holographic film is to be used, the mirror (11), the water cell (5), containing an optically transparent sample (6), a metallised foil target (10) in front of it, power control unit (12) and optical lag line (13,14).

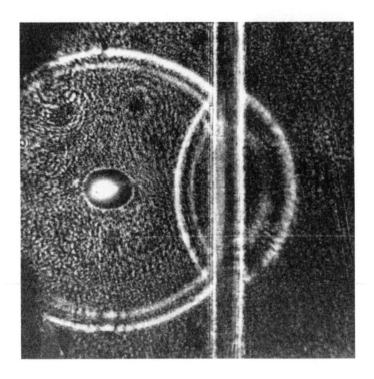

Figure 4.2: Schlieren doubly exposed photograph of an initial shock wave interaction with the PMMA surface.

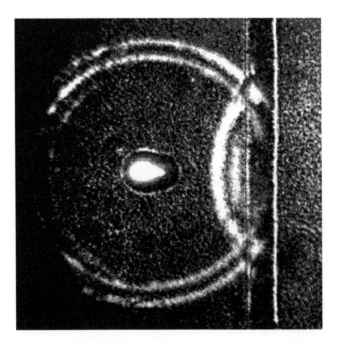

Figure 4.3: Shock interaction with the quartz surface. No longitudinal waves behind an interface.

Figure 4.4: Poisson's waves near an opaque solids.

for ruby laser light with the length $\lambda = 0.69$ μm, SD-3 has the following parameters: $\nu = 0.35$, $\beta = -6 \cdot 10^{10}$ N/m^2, $c_0 = 1.8 \cdot 10^3$ m/s, and $z = 23.0 \times 10^5$ $kg/(m^2s)$for longitudinal waves close to the water wave resistance $z = 15.0 \times 10^5$ $kg/(m^2s)$, the sufficient ratio $E/|\beta| = 0.06$ and comparatively high yield point. The sample in our experiments had the form of straight cylinder of radius $R = 5$ mm with flat length-wise cuts polished for transverse viewing in the transmitted light of ruby laser used for records of reference beams.

The next question to be solved is how to measure the velocity of a longitudinal wave in the rod during the experiment. The wave propagation inside and around the rod was studied (Dreiden et al. 1988, 1989) in order to estimate the difference in wave parameters in solids and liquid. The boundary conical waves were discovered propagating near the lateral surface of the elastic rod loaded by a weak shock wave. These waves were detected using the setup with either shadow photography or holographic registration unit. Observation was done perpendicular to the longitudinal axis of the transparent rod through the polished flat facets made on the lateral surface along the rod and of the opaque tin rod. Doubly exposed shadowgrams and holograms were made in experiments with the same rod. In Figures (4.4, 4.7, 4.6) one can see:

in transparent sample - the wave A having a velocity V_1 in the rod and the decelerating spherical wave B, propagating in the surrounded liquid with the initial velocity V_2. In addition to them, the boundary conical wave P is observed.

in opaque sample (Figure 4.4) only the conical wave P in surrounding liquid is visible using optical setup; the front of it is marked on the photograph with thin black straight line.

The appearance of these waves can be explained as follows: the longitudinal compressive strain wave in the rod is accompanied by the positive transversal displacement V proportional to the Poisson ratio ν, i.e., the rod is swelling locally. This wave has an amplitude proportional to the longitudinal strain and a velocity equal to the wave

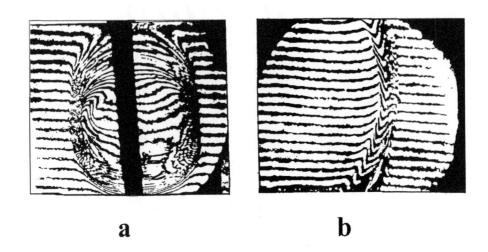

a   **b**

Figure 4.5: Holograms of wave produced by the laser light evaporation of metal layer of the metallised foil, *a*) near the foil; *b*)50 mm from the foil.

velocity in the rod, that provides an appearance of the conical wave in the liquid in the point of intersection of the shock front and the lateral surface, see Dreiden et al. 1986.

According to the Huygens principle the angle α between the front of the conical wave in the liquid and the lateral surface of the rod is governed by:

$$\sin \alpha = V_2/V_1$$

which is in good agreement with direct measurements made using these photographs if we assume both V_2 and V_1 are close to the sound velocities in water ($V_2 = 1465$ m/sec) and in PS ($V_1 = 2350$ m/sec). The essential difference between this conical wave and the well known side wave is that the last is provided by the full internal reflection phenomenon, while the conical wave can be registered along the solid wave guide, arisen by a density wave propagating inside a solid. This conical wave we called the Poisson wave, and is useful to measure the longitudinal wave velocity even in opaque solids.

To make a simultaneous loading of all points of the input cross section of the rod, one has to generate an initial pulse with precisely flat front, otherwise the power of the pulse will be scattered partially due to the full internal reflection. To do so we used the laser beam caused evaporation of a metallic foil immersed into the water cell and separated from the input tip of the rod with water playing a role of an 'acoustical' damper. The holograms of waves in the vicinity of the foil and at the distance of 50 mm from it are shown in Figure 4.5.

The front of the wave near the target is close to absolutely flat, moreover, it is sufficiently flat even far from there. The observation window diameter is 48 mm, providing a good scale for the flat part length estimations. In the consequent experiments the distance between the foil and the rod input was chosen to be equal to $5-7$ mm.

Figure 4.6: Wave pattern at the input of the rod.

Having the flat loading pulse, we began to study the wave picture inside and near the input tip of the cylindrical rod. The photograph made after the reconstruction of the hologram is shown in Figure 4.6. For holographic interferometry a beam splitter 4 divided the laser beam into object and reference beams. The middle cross section of the rod was projected by lens onto the hologram. The first holographic picture was made in the absence of the laser 1 pulse, while the second one was made by a pulse synchronized with a given stage of the wave propagation in the sample 5 inside the water cell 6. The carrier fringes on the interferograms obtained by reconstruction of doubly exposed holograms appear due to rotation of wedge 7 between exposures and show in figures the genuine deformation wave patterns in different areas of the wave guide under study.

In this interferogram the shock wave A moving inside the rod to the right is accompanied by a part of initial pulse D propagating in the surrounding liquid (one can see the lag due to the speed difference). The second shock B entering the rod is formed due to partial reflection of the initial shock from the rod input and the foil consequently. It is to be noted that the thickness of both shocks is much greater than the thickness of the initial compression wave in water. The Poisson waves P are propagating in water and can be easily recognised in the Figure below and above the wave A. Black horizontal stripes represent those parts of the rod that are not transparent: the light cannot propagate through the cylindrical 'caps' along the rod axis, which surface is not perpendicular to the light direction. The vertical black area in Figure 4.6 occured because the input plane and the coherent light direction are not coplanar, hence the tip is not transparent, also.

At the distance of 20 mm from the input the wave pattern is less disturbed (Figure 4.7). Both initial shock A and the shock wave in water B became well separated, the Poisson waves accompany the shock in polystyrene and allow to measure its velocity,

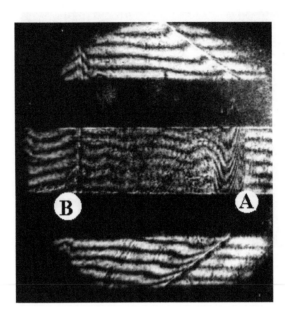

Figure 4.7: The wave pattern in 20 mm distance from the input of polystyrene rod.

while between A and B we can see now a new large wave with the length close to the diameter of the rod.

In order to evaluate properly the holographic interferometry data we should make quantitative estimations of the parameters of the longitudinal wave registered by the optical holography method.

The bulk extension D can be calculated directly in terms of the finite deformation tensor invariants $I_k(\mathbf{C})$ as follows:

$$D = (dV - dv)/dv = \sqrt{I_3(\mathbf{G})} - 1 = \sqrt{1 + 2I_1(\mathbf{C}) + 4I_2(\mathbf{C}) + 8I_3(\mathbf{C})} - 1.$$

After substitutions of $I_k(\mathbf{C}) = I_k(u, v; x, r)$ one has an approximate formula:

$$D \approx (1 - 2\nu)u_x + O(u_x^2).$$

From the other side the conservation of mass yields:

$$\rho_0/\rho = dV/dv \approx 1 + (1 - 2\nu)u_x; \quad \delta\rho = u_x(1 - 2\nu).$$

Otherwise, the relative variation of density is proportional to the transparency indices variation:

$$\delta\rho = -\frac{n_2 - n_1}{n_1 - 1},$$

The new value of the refraction index of the deformed rod n_2 is caused by the local density variation, that can be easily obtained following the Lorenz-Lorentz formula (Born and Wolf, 1964). Then we have

$$n_2 - n_1 = -u_x(1 - 2\nu)(n_1 - 1);$$

where n_1, n_2 are the refraction indices of the material before and after deformation.

The carrier fringe shift caused by the interference is proportional to the difference of the laser light phase variations:

$$\Delta K = \Delta\phi_2 - \Delta\phi_1. \qquad (4.7)$$

Then before the deformation in the reference state we have :

$$\Lambda\Delta\phi_1/(2\pi) = n_0(L - 2h) + 2hn_1;$$

while after it, respectively:

$$\Lambda\Delta\phi_2/(2\pi) = n_0(L - 2h - 2\Delta R) + (2h + 2\Delta R)n_2;$$

where Λ is the light wave length, $2h < R_0$ is rod's thickness along the detecting laser light direction, and the cylindrical cut-offs must be included into consideration.

Finally substituting all estimations,

$$\Lambda\Delta K = 2h(1 - 2\nu)(1 - n_1)u_x + 2\Delta R(n_0 - n_2);$$

and using the relationship $V = -\nu h U_x$ for $r = h$, we obtain for the maximal longitudinal wave (e.g., soliton) amplitude the simple and useful explicit formula:

$$A = \max U_x = -\frac{\lambda\,\Delta k}{2h[(n_1 - 1)(1 - 2\nu) + \nu(n_1 - n_0)]} \qquad (4.8)$$

that allows for calculating the amplitude using the fringe shift data measured directly in photographs.

4.2.1 Strain soliton observation

Next is the step by step detection and observation of the long wave discovered above far from the input cross section, the registration of the wave pattern and its amplitude measurement. The apparatus for excitation and detection of the strain sollton (like that described in Dreiden et al. 1988), was shown in Figure 4.1 .

Solitons were formed in a transparent solid wave guide from a primary shock wave generated in a liquid near the end of a rod by laser vaporization of a metal target (Figure 4.5). The power density of the laser radiation acting on the target was monitored by the energy measuring device and was held constant at the level of $2.3 \cdot 10^8 W/cm^2$ for the entire experiment.

After the pulse-solids interaction's study we have made some improvements in the experimental method that enhanced the capability of detecting the waves. To observe the evolution of a wave we used the 149 mm long polystyrene rod and the field of view for taking the holographic interferograms of the wave was increased to 50 mm. The longitudinal strain wave was observed at different distances from the input tip of the rod by shifting the cell with the rod relative to the beam for recording the hologram. It allowed us to obtain an image of the entire wave pattern in the rod in a single

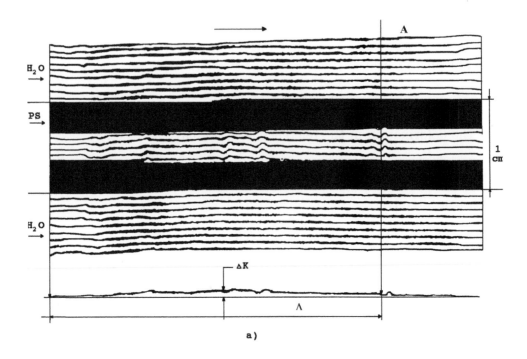

Figure 4.8: Strain soliton in the interval of PS rod of 40-90 mm from the input tip.

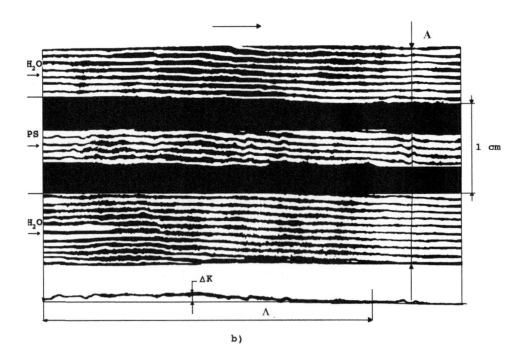

Figure 4.9: The same soliton at the interval of 60-110 mm from the input. The distance between shock wave A and the soliton increases.

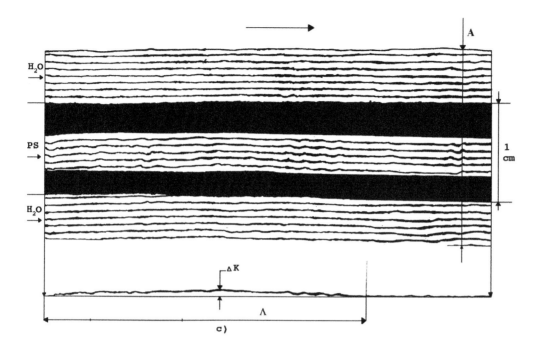

Figure 4.10: The strain soliton remains of permanent shape at the interval of 75-125 mm from the input tip, while the shock wave A almost disappeared.

frame, and to observe the disturbances at three different areas of the rod: between 40 and 90 mm (Figure 4.8), between 60 and 110 mm (Figure 4.9), and between 75 and 125 mm (Figure 4.10) from the input tip of the rod.

From these interferograms one can see in the PS rod the primary shock wave A that initiated the compression bulk soliton shown in the auxiliary footnote graph under each interferogram for convenience. The secondary shock wave of a complex shape (the sharp small kink in the fringe in the middle of the soliton) is visible, which forms as a result of repeated reflections of the primary shock wave in the liquid from the end of the rod and then from the target and passes through the rod. As it can be seen in these three figures the shock wave damps out farther along the rod but their velocity remains higher than the sound speed in polystyrene and higher than the velocity of the soliton (the soliton is displaced further from the shock waves from frame to frame). The soliton turns out to be a fairly long trough-shaped longitudinal density wave outside of which there is no rarefaction wave of any appreciable amplitude. In the liquid surrounding the rod the interference pattern remains horizontal, i.e., undisturbed, which confirms that the wave that is measured propagates in the rod. The dynamics of the motion of the strain soliton at various distances clearly shows that the shape of the wave changes very little, which is an essential property of a solitary nonlinear wave of longitudinal strain.

During these experiments the second ruby laser 2 has been used for the hologram recording, according to the scheme shown in Figure 4.1. The shutters of both lasers 1 and 2 were synchronized by a multichannel generator of delayed pulses that allowed the recording of a wave pattern at a required time moment with the accuracy of order

$0.5 \cdot 10^{-6}$ seconds. The light beam from the laser 2 (the beam diameter was 1.5 mm) was expanded by a telescopic system 3 up to the diameter equal to 50 mm, and then it was divided into the object and reference beams by a wedge 4. Passed through the wedge, the object beam was directed to the water cell 5 and to the rod 6 immersed into it. The central rod section was projected onto the hologram plane 9 by a lens 8, and the hologram of the focused image was recorded.

The first exposure of the hologram was carried out by the pulse from the laser 2 in the absence of a pulse from the laser 1, so the hologram of undisturbed wave guide 6 was recorded. The second exposure was made by a laser pulse synchronized with the prescribed stage of the wave propagation. Observations were made in the transversal direction. The carrier fringes on interferograms, obtained due to the reconstruction of doubly exposed holograms, occured due to the wedge 7 turn between the exposures. The longitudinal strain wave patterns were recorded at various distances from the input edge of the rod that was attained by the cell displacement along the axis of wave propagation. Recently Harith et al. (1989) reported about the study of a shock wave produced in water by the laser explosive evaporation of a metallic film target immersed into water. It was shown that such a shock wave exhibits a very narrow compressive area (0.1-0.2 μm wide) followed by a considerable rarefaction area (1 mm wide) of small amplitude. The parameters of this shock wave satisfy the conditions required for the strain soliton generation.

The soliton parameters calculations were based on the data of the holographic interferograms obtained. Note that the interferometric pattern does not exhibit a standard bell-shaped image of a shallow water soliton since the strain soliton is, in fact, the *longitudinal* long density wave in a solid.

The soliton amplitude is calculated using the interference fringe shift ΔK measured in the interferogramme according to (4.8). The amplitude is determined by the maximal fringe shift value. Derivation of the amplitude formula shows that the length Λ of the elastic strain solitary pulse may be directly determined from the interferogram as the length of the fringe perturbation between two undisturbed areas.

For the experiments under consideration the parameter values are the following $n_0 = 1.33, n_1 = 1.6, 2h = 7.5$ mm and $\lambda = 0.69 \cdot 10^{-7}$ m. Calculation of the parameters of the compressive strain soliton whose interferogram is shown in Figure 4.8 gives the following values of the amplitude $A = 3.29 \cdot 10^{-4}$ and the wave length $\Lambda = 32.5$ mm. An analysis of the interferograms shown in Figures 4.9, 4.10 gives essentially the same values for the propagating wave. The characteristic lengths of the solitary waves as calculated by the DDE theory (2.32) and the refined DDE theory (2.54) using the experimentally determined amplitudes A are equal, respectively, to 44.9 and 57.1 mm. Therefore, the theoretically predicted values do not exceed the experimental ones by more than 38% and 75% respectively. The remaining discrepancy can be explained by the fact that the soliton of the longitudinal strain is the long wave with very smooth fronts. It is thus rather difficult to make a more accurate measurement of the pulse width (the wave length) Λ, although the values that we obtained seem to be satisfactory within the limits of accuracy of measuring the amplitude of the soliton.

4.2.2 Why is it a soliton?

Note that in any case the soliton has the length greatly (up to 7 times) exceeding the radius of the wave guide and does neither disperse nor dissipate even at the long distance (up to 60 radius units) from the excitation area.

In summary, we may confirm that we have developed both the theory and the method of experimenral investigation of the bulk solitary waves of longitudinal strain and hence are able to:

- observe a solitary strain wave over its entire length;

- obtain a picture of its evolution along the rod and thereby demonstrate experimentally that this wave does indeed preserve its shape and amplitude/length ratio, which identifies it as the genuine longitudinal strain soliton;

- calculate and measure the wave parameters.

The following arguments may confirm the observation of the genuine strain solitary wave in our experiments.

Firstly, there is *no tensile area behind* the observed long compressive wave (having a length $\Lambda > 7\ R$), that is the distinctive feature of the localized nonlinear waves. Tensile areas, if any, can be easily detected using the same apparatus: the fringes will be shifted in an opposite direction. However, deformation of the rod behind the soliton was studied in detail, and nothing was observed there except straight fringes, i.e., the rod was free of strain again after the soliton propagation.

Secondly, even at the distances, exceeding dozens of rod's radii, both *the shape and the wave parameters remain permanent* and do not exhibit any essential distortions in the uniform rod, as it was shown recently (Dreiden et al. 1995), that is, the nonlinear strain wave possesses another distinctive feature of a soliton. The distance chosen for our observations seems to be sufficiently large, because polystyrene is well known to be an effective absorber of acoustic and shock waves. The last is confirmed by considerable decay and even disappearance of the shock wave that moved ahead of the soliton, as can be seen in Figure 4.10.

One of the most important results for physics and technology consists in possible transportation of the elastic deformation energy using solitons at long distances without losses even in materials, having a considerable absorption (dissipation) even for shocks. We proved in experiments that the elastic strain soliton is not absorbed even at the distances much greater than standard linear dissipation length for polystyrene. Presumably it means that the nonlinear absorption is much less important than the linear one that does not affect the soliton due to the difference in wave velocities, however this problem requires further analysis.

The measurement of the wave amplitude is supposed to be quite plausible for the comparison with the theory. One can see in Figures 4.9, 4.10 that the maximal amplitude of the strain soliton is achieved at the distances 60 and 95 mm from rod's input edge, respectively.

However, some new theoretical results cannot be checked in our experiments and require further study, namely:

i) the refined theory was proposed to describe the nonlinear strain waves in a rod, improved by means of better fulfilment of the boundary conditions on a free lateral rod surface. The differences in the dispersive terms coefficients values in (2.32, 2.54) result in variation of the value of the soliton width Λ. The calculations based on the experimental data for the homogeneous rod show the 20-25% alteration in its value with respect to those found on the basis of previous theory (Samsonov, 1984; 1988a). However, the concept of a soliton width is conventional, therefore this particular deviation between two theories is rather difficult to confirm in our experiments.

ii) The upper 'speed limit' was found in (4.4) for velocities of a soliton in any elastic material, and the 'allowed zone' for velocities was found to exist, outside of which neither generation nor propagation of a soliton are possible. The upper value of the limit in (4.4) was not proven in our experiments because the soliton amplitude corresponding to this velocity provides an inelastic strain in polystyrene.

iii) The observation of the amplitude dependence upon the Poisson coefficient ν is expected to be of interest for applications found for the first time by means of the exact formula (4.8).

The theory developed here is of importance for tentative experimental study of the periodical, particularly cnoidal waves, because the deviation in the values of the wave length can be measured in experiments with reasonable accuracy.

4.2.3   Reflection of a strain soliton

There are no infinite wave guides in practice of technology; rods and plates and shells are combined in structures, and the problem of reflection of the nonlinear strain wave is to be considered under different conditions of clamping of the finite wave guide end.

We shall use the standard approach based on the Hamilton principle requiring the action functional variation vanishes

$$\delta S = \delta \int_{t_0}^{t_1} dt \left[2\pi \int_{-\infty}^{X} dx \int_{0}^{R} r\, \mathcal{L} dr + A \right] = 0, \qquad (4.9)$$

however A notes now the work of an external force at the end of the rod: $x = X$.

Assuming that the Murnaghan model is valid for the elasticity of the rod, and the lateral surface of the rod is free of stresses, we obtain the DDE (2.54) for the strain component $u = U_x$ in a form:

$$u_{tt} - \frac{E}{\rho} u_{xx} - \left(\frac{\beta}{2\rho} (u^2) + aR^4 u_{tt} - bR^2 u_{xx} \right)_{xx} = 0, \qquad (4.10)$$

where $a = -(\nu(1-\nu))/2, \quad b = -\nu E/(2\rho)$. Moreover the Hamilton principle requires that for $x \to -\infty$ the values of u and its derivatives in x and t should vanish. At the other end $x = X$ the boundary conditions depend on the type of clamping.

For free end we have $A = 0$ and from (4.9) it yields:

$$u = 0, u_{xx} = 0. \qquad (4.11)$$

When the end is clamped, the external force work is not determined and the kinematic conditions of zero displacement and its velocity are to be given:

$$U = 0, U_t = 0, \qquad (4.12)$$

or in terms of strains it yields:

$$u_x = 0, \; u_{xt} = 0. \tag{4.13}$$

Introducing the scales as before: Λ for x, Λ/c_0 for t, A/Λ for u, where $c_0 = \sqrt{E/\rho}$, we can write the dimensionless version of the DDE for further asymptotic analysis. The small parameter ε should be chosen to balance the nonlinearity and dispersion

$$\varepsilon \; = \; A \; = (R/\Lambda)^2 \; << \; 1. \tag{4.14}$$

Assume u depends on x, t and also upon the slow time variable $\tau = \varepsilon t$. Finding the solution to (4.10) in power series in ε:

$$u = u_0 + \varepsilon u_1 + \ldots . \tag{4.15}$$

substituting it into the DDE, we obtain the D'Alembert solution for u_0:

$$u_0 = u_{01}(\xi, \tau) + u_{02}(\eta, \tau), \tag{4.16}$$

where $\xi = x + t$, $\eta = x - t$. In order of ε the equation is the following:

$$2u_{1,\xi\eta} = 2u_{01,\xi\tau} - 2u_{02,\eta\tau} + \frac{\beta}{2E}\left((u_{01}^2)_{\xi\xi} + 2v_{01,\xi}u_{02,\eta} + (u_{02}^2)_{\eta\eta}\right) + \frac{\nu^2}{2}(u_{01,4\xi} + u_{02,4\eta}) \tag{4.17}$$

and the necessary condition of absence of secular terms leads to two uncoupled KdV equations for u_{01} and u_{02},

$$2u_{01,\tau} - \frac{\beta}{2E}(u_{01}^2)_\xi - \frac{\nu^2}{2}u_{01,\xi\xi\xi} = 0, \tag{4.18}$$

$$2u_{02,\tau} + \frac{\beta}{2E}(u_{02}^2)_\eta + \frac{\nu^2}{2}u_{02,\eta\eta\eta} = 0, \tag{4.19}$$

while the 1st order problem solution is (4.17):

$$u_1 = \frac{\beta}{E}u_{01}u_{02} + u_{11}(\xi, \tau) + u_{12}(\eta, \tau). \tag{4.20}$$

Substituting the soliton solutions of the KdV equations into the leading order solution, we obtain in original notation

$$u_0 = \frac{6E\nu^2}{\beta}k^2\left[\cosh^{-2}\left(k[x + (1 + \varepsilon\nu^2 k^2)t - x_{01}\right) + \cosh^{-2}\left(k[x - (1 + \varepsilon\nu^2 k^2)t - x_{02}]\right)\right]. \tag{4.21}$$

where x_{0i} are constant phase shifts.

Therefore the type of the strain wave depends only on sgnβ and not on the direction of motion. Both conditions (4.13) on the clamped end will be satisfied if $x_{02} = 2X - x_{01}$, and the solitary strain wave will be reflected without shape change. On the contrary, the solitary wave (4.21) does not satisfy the free end conditions (4.11), and will disappear due to either dispersion or unbalanced nonlinearity action. The IST method may be used for analytical study of distortion of a solitary wave, see Ablowitz, Segur (1981), Dodd et al. (1984).

Another way to study the reflection process was in the numerical simulation. Following Dreiden et al. (2001), we describe briefly the main differences of it from the standard Godunov implicit finite difference scheme which will be used as the general tool for numerical simulation approach in Chapter 6.

Boundary conditions at the end of the rod may be modelled by symmetric prolongation of the calculation interval after the real end, i.e., the simulation will be done in $0 < x < 2X$. For a free end the finite difference version of the DDE (2.54) is solved in the interval $X < x < 2X$ with sgnβ opposite to that used in $0 < x < X$. Initial pulses are assumed to be solitons located centrally symmetric with respect to the genuine end $x = X$ of the rod. The type of initial soliton depends on sgn β, e.g., for $\beta > 0$ in $0 < x < 2X$ the soliton should be a tensile wave, while the compression soliton is in the interval $X < x < 2X$.

Initial velocities of solitons are to be given equal and to meet each other. The boundary condition $u(x = X) = 0$ is used explicitly, while the second order derivative is to be zero automatically in calculations.

Reflection from a free end $0 < x < X$ shown that the amplitude of moving wave decayed at the free end, reflected wave was of opposite polarity and disappeared due to dispersion. Eventually there is no localized strain wave at the input tip of the rod.

For the clamped end simulation in the region $0 < x < 2X$ calculations were subjected to the axial symmetry with respect to the line $x = X$. In both subdomains the similar DDE is solved and initial pulses are assumed to be located on equal distances from the line. The boundary conditions (4.13) are satisfied automatically. In accordance with the theory the doubling of the amplitude of initial soliton occured at the end, the reflected wave had the same sign, amplitude and velocity and propagated to the input without decay.

Now we will discuss the physical experiments in strain soliton reflection that were made using the apparatus described above.

We have used for reflection experiments the solitary strain wave generated in the polystyrene SD-3 rod by means of the experimental the setup with the parameters: $n_0 = 1.33$, $n_1 = 1.6$, $\lambda = 0.69 \cdot 10^{-6}$ m. Polystyrene (PS) has $\nu = 0.35$ and the negative nonlinearity coefficient β, hence the upward fringe shift in the holographic interferogram corresponds to the compression deformation.

Reflection from the free end is shown in Figure 4.11. The soliton is moving to the right in all photographs (except the last one in this subsection), and the fringes are not disturbed in water behind the end of the rod, that confirms the whole reflection of the falling wave. In clear agreement with the theory in the footnote in Figure 4.11-b, one can see the decrease of the amplitude of compression wave in comparison with the value of the initial one in Figure 4.11-a. There is no any localized wave in the entire end of the rod, which confirms experimentally the theoretical prediction of total absence of tensile solitary wave in polystyrene.

To model the clamped edge of the rod we glued it with different massive solid plates. First, we studied the penetration of the wave through the clamped end into

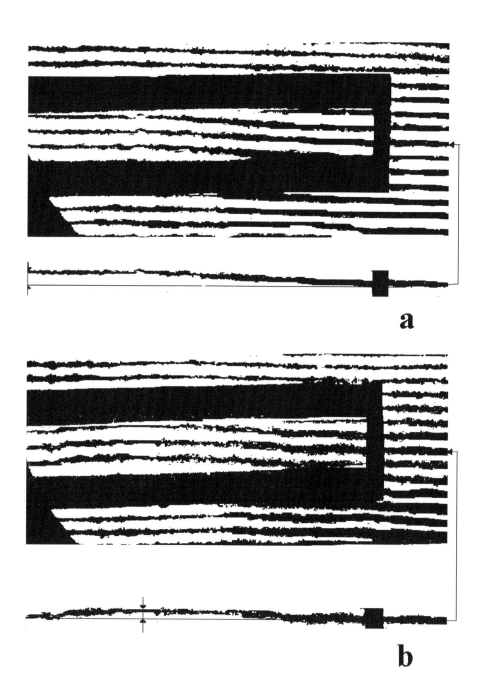

Figure 4.11: Experiments in decay of the soliton, reflecting from a free end.

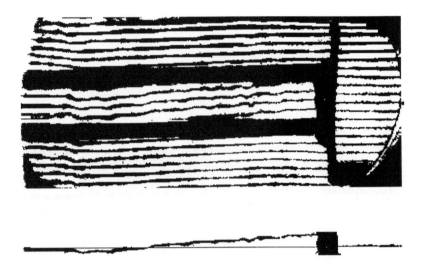

Figure 4.12: Total reflection of the strain wave in the PS rod glued with the quarz plate (at the right end).

the transparent solids to visualise the wave propagation process. It is shown in Figure 4.12 that the compression wave cannot propagate through the PS-quartz interface (the acoustical resistance z of quartz is 6 times more than of PS), while it may easily move through the interface PS-PMMA (values of z of both materials are close) as it can be seen in Figure 4.13. Acoustical resistance of brass is 15 times more that of PS, therefore the elastic wave energy should reflect almost totally from the rod end clamped with brass.

In Figure 4.14 one can see the strain soliton moving from the left and reflecting from the brass plate, clamping the end of the rod. The plate is located to the right and shown in figures as the black vertical strip. The amplitude of the soliton is almost doubled (Figure 4.14-b) during the reflection, and returns to the initial value after it, when soliton is moving *to the left* and even far (140 mm) from the clamped end (Figure 4.15-c). Note that the surrounding water is disturbed, therefore to estimate the amplitude correctly one should subtract the fringe shift outside rod (fringe in the footnote B) from the shift inside the rod (footnote A). It results in almost equal value of the soliton amplitude to that observed in Figure 4.14-a.

Therefore we have shown that the strain solitary wave reflection from the free end of a wave guide transforms the soliton to the opposite sign wave, which disappears shortly, while being reflected from the clamped end the soliton propagates along the wave guide with permanent shape and velocity. It fits well the theory of reflection and even of collision of solitary waves and may provide a new experiment in head-on collision of strain solitons in solids. Moreover the DDE is proved to be valid for solitary wave propagation description in finite length wave guides. The wave observed after reflection possesses the main feature of the soliton to keep its shape in collision, therefore we confirmed this property of elastic solitons in experiments.

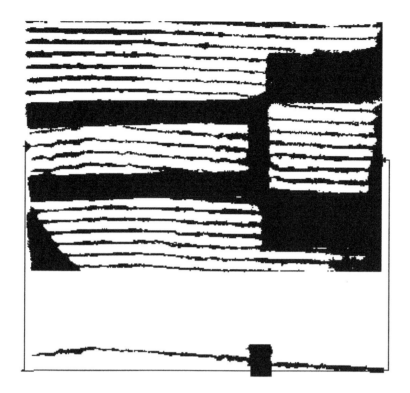

Figure 4.13: Penetration of the long strain wave, moving to the right, through the PS-PMMA interface.

4.3 Solitons in inhomogeneous rods

Possible inhomogeneity of a wave guide can be intentionally designed and/or resulted from imperfections during manufacturing. It can influence the wave propagation strongly, even being slow with respect to the appropriate scale, and it will be studied thoroughly for this reason.

This section is devoted to the theoretical and experimental description of the propagation and amplification of the strain solitary wave (soliton) in a cylindrical nonlinearly elastic rod with variations in cross section area or in elasticity. We call it an inhomogeneous rod in the following for convenience, while the uniform rod will be called the homogeneous one.

The soliton dynamics in the non-uniform rod was firstly studied analytically in (Samsonov, 1982;1984), based on the reduction of the perturbed DDE to the perturbed KdV equation. Limitations of it are evident, however most of the qualitative estimations remain valid, being reconsidered on the basis of the DDE approach. The main difference is that the elastic solitary wave does not possess the wave collapse as it seemed to be in the framework of the KdV approach. Moreover, we note that the *adiabatic* approach to solve the perturbed KdV proposed by Ko and Kuehl (1978), was insufficient to describe the nonuniform dependence of the inhomogeneity parameter δ in a solution, see Karpman and Maslov (1978) for further details.

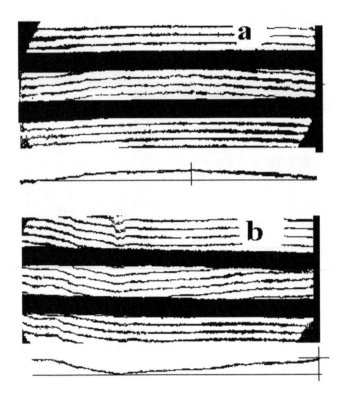

Figure 4.14: Experiments in reflection of the strain soliton from the clamped end.

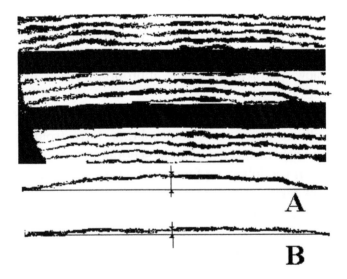

Figure 4.15: Soliton moves *to the left* after reflection and keeps the permanent shape. The photograph was made far (140 mm) from the clamped end.

The wave propagation problem for the physically inhomogeneous rod was considered and the solution was found in (Samsonov, Sokurinskaya, (1987)), where the evolution of longitudinal displacement velocity soliton was described by the perturbed KdV equation and solved by means of perturbation technique, see (Kodama, Ablowitz, 1981). We have shown the limitations of the KdV approach in nonlinear wave problems in solids and now we shall apply the model based on the DDE (2.33) and (2.54) having the variable coefficients.

When the rod is *inhomogeneous* and uniform $R(x) = R_0 - const$, the equation (2.29) was derived written in terms of the strain component $u = U_x$:

$$u_{tt} - \frac{\partial}{\partial x}\left[\frac{1}{\rho}(Eu)_x\right] = \frac{\partial}{\partial x}\frac{1}{2\rho}\frac{\partial}{\partial x}\left[\beta u^2 + \rho\nu^2 R_0^2 u_{tt} - R_0^2\frac{\partial}{\partial x}\left(\mu\nu^2 u_x\right)\right]. \qquad (4.22)$$

When the rod is *non-uniform* but the homogeneous one, the problem may be governed by the equation:

$$u_{tt} - c_0^2\frac{\partial}{\partial x}\left[\frac{1}{R^2}\frac{\partial}{\partial x}\left(R^2 u\right)\right] = \frac{\partial}{\partial x}\frac{1}{2R^2}\frac{\partial}{\partial x}\left[\frac{\beta}{\rho}R^2 u^2 + \nu^2 R^4 u_{tt} - c_1^2\nu^2\frac{\partial}{\partial x}\left(R^4 u_x\right)\right].$$
$$(4.23)$$

followed from (2.30); it differs from the previous one, in particular, by the constant value of the linear wave velocity. The refined model used for the DDE improvement will lead to corresponding changes in coefficients that may be taken into consideration easily.

4.3.1 Solitons in the non-uniform rod

Firstly we shall consider the problem on the influence of the small non-uniformity of a wave guide on the solitary strain wave moving from the uniform part of the wave guide, where it was generated, into the non-uniform one. We assume the slow dependence on x for the parameters, consider the geometrically nonuniform rod only with varying cross section area: $S = S(\delta x), \delta << 1$ and perform the formal asymptotic analysis of the problem.

After transformation of the DDE to the dimensionless form we obtain the p.d.e. of the form

$$4u_{tt} = \frac{\partial}{\partial x}\left\{\frac{1}{S}\frac{\partial}{\partial x}\left[4uS + 6u^2 S + S^2(au_{tt} - bu_{xx}) - 2bSS_x u_x\right]\right\}, \qquad (4.24)$$

that can be written as

$$4u_{tt} = F_{xx}^0 + \frac{\partial}{\partial x}\left[F^0 S_x/S\right], \qquad (4.25)$$

where

$$F^0 \equiv 4u + 6u^2 + S(au_{tt} - bu_{xx}) - 2bS_x u_x \qquad . \qquad (4.26)$$

Using the asymptotic expansion approach, we introduce the slow variable $X = x\delta$, and define an unknown phase variable ϑ by the space-similar transformation:

$$\begin{cases} \vartheta_t = -1 \\ \vartheta_x = A(X) \end{cases} \tag{4.27}$$

where $A(X)$ is a new unknown function of slow variable X. Hence we can calculate

$$\begin{aligned} S_x &= \delta S_X; \\ u_{tt} &= u_{\vartheta\vartheta}; \quad u_x = Au_\vartheta + \delta u_x; \\ u_{xx} &= A^2 u_{\vartheta\vartheta} + \delta(2Au_{\vartheta X} + A'u_\vartheta) + O(\delta^2), \end{aligned}$$

and expanding the strain u in power series with respect to δ, in the form $u = \sum_{i=} \delta^i u^{(i)}$, we obtain from (4.25)

$$(4u^0 - A^2 F^0)_{\vartheta\vartheta} = \delta\left\{ \frac{(AS)_X}{S} F_\vartheta + 2AF^0_{\vartheta X} - b\left[SA^3(\log S^2 A)_\vartheta u_{\vartheta\vartheta\vartheta} + 2S^3 u_{\vartheta\vartheta\vartheta X} \right] \right\}$$
$$+ O(\delta^2)$$

where $F^0 \equiv 4u^0 + 6(u^0)^2 + S(a - bA^2)u^0_{\vartheta\vartheta}$, and

$$F \equiv 4u + 6u^2 + S(a - bA^2)u_{zz} \propto F^0 + \delta\left[4u^1 + 12u^0 u^1 = (a - bA^2)Su^1_{\vartheta\vartheta} \right] + O(\delta^2).$$

The zero-order problem takes the form

$$\begin{aligned} (4u^0 - A^2 F^0)_{zz} &= 0, \\ (\partial^k u^0 / \partial\vartheta^k) &\to 0 \text{ for } |\vartheta| \to \infty, \end{aligned} \tag{4.28}$$

which possesses the soliton solution in the form:

$$u^0 = \alpha \cosh^{-2} \eta(\vartheta - \vartheta_0), \tag{4.29}$$

where $\alpha \equiv (1-A^2)/A^2$, the inverse wave length η defined now as $\eta^2 = [1-A(X)]/[SA^2(a-bA^2)]$, and the phase shift parameter ϑ_0 all depend upon the unknown $A(X)$. Note that it follows from (4.28) that $F^0 = 4u^0/A^2$. We arrive after a little algebra and introduction of the new function of ϑ and X as $\varphi \equiv \eta(\vartheta - \vartheta_0)$, so that:

$$\frac{\partial u^0}{\partial X} = u^0 \frac{\partial}{\partial X}(\log\alpha) + \varphi\frac{\partial u^0}{\partial\varphi}\frac{\partial}{\partial X}(\log\eta) - \eta\frac{\partial u^0}{\partial\varphi}\frac{\partial\vartheta_0}{\partial X},$$

and

$$\frac{\partial F^0}{\partial X} = 4(1 + \alpha)\frac{\partial u^0}{\partial X} + 4u^0 \frac{\partial\alpha}{\partial X},$$

at the first order problem for u^1. It is, evidently, the linear one, and takes the form

$$\left[4(1 - A^2)u^1_{\vartheta\vartheta} - A^2(12u^0u^1 + (a - bA^2)Su^1_{\vartheta\vartheta}) \right]_{\vartheta\vartheta}$$

$$= \left[F^0(SA)_X/S + 2AF^0_X - 2bSA^3u^0_{\vartheta\vartheta X} - bSA^3u^0_{\vartheta\vartheta}(\log S^2 A)'_X \right]'_\vartheta \quad (4.30)$$

Then reducing the order of the equation further by means of integration it with respect to φ, with the boundary conditions of vanishing of u^0 as $|\varphi| \to \infty$ being taken into account to calculate auxiliary integrals, we obtain the inhomogeneous Legendre equation for $u^1(\varphi)$ in the form

$$\frac{\partial^2}{\partial\varphi^2}u^1 - 4(1 - 3\cosh^{-2}\varphi)u^1 = C_1(1 - \tanh\varphi) + C_2\varphi u^0 + C_3u^0\frac{\partial\vartheta_0}{\partial X} +$$

$$bSA\eta\alpha^{-1}\left(C_4\varphi u^0_{\varphi\varphi} + C_5u^0_\varphi + C_6u^0_{\varphi\varphi}\right) \quad (4.31)$$

where all C_i are the functions of the slow variable, as follows:

$$C_1 = \frac{4}{A^3\eta}\frac{\partial}{\partial X}(\log\alpha^2 SA^{-3}\eta^{-2}), C_2 = -\frac{8}{A^3\eta}\frac{\partial}{\partial X}(\log\eta),$$

$$C_3 = \frac{8}{A^3\eta}, C_4 = 2\frac{\partial}{\partial X}(\log\eta), C_5 = \frac{\partial}{\partial X}(\log AS^2\alpha^2\eta^2), C_6 = -2\eta\frac{\partial\vartheta_0}{\partial X}.$$

Let us denote the right hand side in (4.17) as $\Phi(u^0)$. To exclude the secular terms in the asymptotic solution it is necessary to satisfy an orthogonality condition

$$\int_{-\infty}^{\infty} u^0\Phi(u^0)d\eta = 0, \quad (4.32)$$

that leads to an equation for the unknown $\alpha(X)$:

$$\frac{\partial}{\partial X}\log\frac{S\alpha^2}{A^3\eta} + \frac{bSA^4\eta^2}{5(1+\alpha)^2}\frac{\partial}{\partial X}\log S\alpha^2 A\eta = 0,$$

where $A = A(\alpha(X)), \eta = \eta(\alpha(X))$, and the function $S = S(X)$ is given. The solution of it will be written below and analysed in the dimensional form, here we indicate only the dependence of the solitary wave amplitude $\alpha(X)$ upon the value of cross section area $S(X)$, namely, any decrease of $S(X)$ along the rod axis results in a noticeable increase of the amplitude (amplification) and, thus, in focusing of the impulse, and vice versa.

The first order problem solution u^1 of the inhomogeneous Legendre equation, bounded for $|\varphi| \to \infty$, takes the form

$$u^1 = -\frac{C_1}{4}\tanh^2\varphi + \left[\frac{C_2}{8} - \frac{3b}{5S\alpha^2}\frac{\partial}{\partial X}(A\eta S^2\alpha^2)\right]\tanh\varphi + F(\varphi), \quad (4.33)$$

where

$$
\begin{aligned}
F(\varphi) \;=\; & 15D\cosh^{-2}\varphi\tanh\varphi + 2\vartheta_{0,X}\,(1-\varphi\tanh\varphi)/(\alpha A) - \\
& 3(\log\ SA^{-3}\alpha^2\eta^{-2})'_X(\varphi\tanh\varphi\cosh^{-2}\varphi)/(\eta A^3) + \\
& +2(\log\ SA^{-3}\alpha^2\eta^{-2})'_X(2-3\tanh^2\varphi) - \eta_X u^0\varphi(2-\varphi\tanh\varphi)/(\alpha A^3\eta) + \\
& bSA\eta(2\eta\vartheta_{0,X} + \varphi(\log\ S^2A\alpha^2\eta)'_X\varphi\tanh\varphi\cosh^{-2}\varphi),
\end{aligned}
$$

and D is the constant of integration, whence $F \to 0$ for $|\varphi| \to \infty$ and $F \neq 0$ for $\varphi = 0$. The height h of a shelf or a plateau behind the soliton, that is in fact a small wave of strain, very long even in comparison with the width of the soliton, propagating behind a pulse, is determined by the limit

$$
h = \lim_{\varphi\to-\infty}(u^0+\delta u^1) = -\delta(1+\alpha)(S\frac{\sqrt{a-b+\alpha a}}{\alpha})\frac{d}{dX}(\log[\alpha^5(1+\alpha^3)(a-b+\alpha a)^3 S^7],
$$

and is proportional to the derivative $(-dS/dX)$. When $S_X < 0$, an impulse is self-focusing , and the plateau of positive height is the long compression strain wave with a velocity considerably smaller the main velocity V of a soliton u^0. From the physical viewpoint this qualitative result is similar to the one obtained formerly in (Samsonov, 1984) in the framework of the KdV approach. The essential difference is that a critical value of S, responsible for the formally unbounded increase of amplitude of the KdV soliton, does not exist, when the DDE based model is considered.

To complete the solution of the first order problem it is necessary to obtain a solution uniformly valid in ϑ by means of the matched asymptotic expansion method. This solution contains a region of non-uniformity ahead of a soliton, a quasi stationary solution $u^0 + \delta u^1$ with variable parameters and an oscillating tail behind the shelf. It could be done in a routine manner, however it is a complicated task, and we shall study these waves numerically below and find the space and time behaviour of the strain soliton propagating in non-uniform rod.

Evidently, the localized wave evolution in a sharply inhomogeneous rod, when the hyperbolic wave operator will not be the fundamental one and cannot be covered by means of this consideration. It will require the numerical simulations presented in Chapter 6.

4.3.2 Solitons in the inhomogeneous rod

The wave guide may be physically inhomogeneous, i.e. the *2nd* order elastic moduli is assumed to vary along the x-axis also and slowly enough to be modelled by means of the dependence on small parameter $\delta \ll 1$ as $E = E(\delta x), \nu = \nu(\delta x)$, while the *3d* order moduli variations along the rod axis, e.g., of the Murnaghan material, are described in general by the dependence of the nonlinearity coefficient on x as $\beta = \beta(E,\mu;l,m,n) = \beta(\delta x), \delta \ll 1$. For simplicity and to extract the influence of elasticity variations, the rod's cross section is supposed to be uniform now.

After transformation of the perturbed DDE to dimensionless form in a way similar to the previous one used to solve the wave problem in the nonuniform rod we get

$$4u_{tt} = \frac{\partial}{\partial x} \left\{ \frac{1}{\rho} \frac{\partial}{\partial x} \left[4Eu + 6\beta u^2 + \rho \nu^2 u_{tt} - b\frac{\partial}{\partial x}(\mu \nu^2 u_x) \right] \right\}. \tag{4.34}$$

Again, the perturbation appeared in the main hyperbolic (wave) operator, but for the following transformations, it is convenient to rewrite the equation in a different way as

$$4u_{tt} = F_{xx} + \frac{\partial}{\partial x} \left[\frac{F\rho_x}{\rho} \right],$$

where

$$F \equiv 4Eu/\rho + 6\beta u^2/\rho + \nu^2 u_{tt} - (b/\rho)(\mu \nu^2 u_x)_x$$

Introduce slow variable $X = x\delta$, the phase variable ϑ, using the unknown function $A(X)$ and two coupled differential equations (4.27), and writing the moduli derivatives in new variables we get

$$E_x = \delta E_X; \nu_x = \delta \nu_X;$$

while derivatives of the unknown function are expressed now as before:

$$u_{tt} = u_{\vartheta\vartheta}; u_x = Au_\vartheta + \delta u_x; u_{xx} = A^2 u_{\vartheta\vartheta} + \delta(2Au_{\vartheta X} + A' u_\vartheta) + O(\delta^2).$$

We are dealing with the DDE, having a solitary wave solution in absence of perturbations, therefore any expansion with respect to the small parameter ϵ is not necessary. Expanding the unknown strain function $u(\vartheta, X)$ in power series with respect to the only small parameter δ of inhomogeneity as $u = \sum_{i=} \delta^i u^{(i)}$, we have

$$(4u - A^2 F)_{\vartheta\vartheta} = \delta \left\{ \frac{(A\rho)_X}{\rho} F_\vartheta^0 + 2A\frac{\partial^2 F^0}{\partial X \partial \vartheta} - b\frac{A^2}{\rho} \left[(A\mu\nu^2)_\vartheta u_{\vartheta\vartheta\vartheta} + 2A\mu\nu^2 u_{\vartheta\vartheta\vartheta X} \right] \right\} \tag{4.35}$$

where the following function

$$F \equiv F^0 + \delta \left[4 \left(1 - \frac{E_0 A^2}{\rho_0} \right) u^1 + \frac{12\beta_0}{\rho_0} u^0 u^1 + A^2 \nu_0^2 \left(1 - \frac{bA^2\mu_0}{\rho_0} \right) u_{\vartheta\vartheta}^1 \right] + O(\delta^2).$$

was introduced to detach explicitly the dependence of δ and of the unperturbed values, marked here and below by zero indices, and

$$F^0 \equiv \frac{4Eu^0}{\rho_0} + \frac{6\beta_0(u^0)^2}{\rho_0} + \nu^2 \left(1 - \frac{b\mu_0 A^2}{\rho_0} \right) u_{\vartheta\vartheta}^0,$$

As before, the solution of zero order problem written now as

$$(4u^0 - A^2 F^0)_{\vartheta\vartheta} = 0,$$

with the boundary conditions $(\partial^k u^0 / \partial \vartheta^k) \to 0$ for $| \vartheta | \to \infty$, has the form of the solitary wave:

$$u^0 = \alpha \cosh^{-2} \eta(\vartheta - \vartheta_0), \tag{4.36}$$

where the amplitude α is defined by means of unperturbed parameters and an unknown function $A(X)$, according to the expression

$$\alpha \equiv \frac{1}{A^2} \left(\frac{\rho_0}{\beta_0} - \frac{A^2 E_0}{\beta_0} \right)$$

and $\eta^{-1} = \eta^{-1}[A(X)]$ is the wave length. Moreover, it is easy to calculate from the zero order equation that $F^0 = 4u^0/A^2$, due to the prescribed boundary conditions.

The first order problem is governed by the 4th order o.d.e., that is linear with respect to u^1. It can be written with coefficients, depending upon the slow variable X:

$$\frac{\partial^2}{\partial \vartheta^2} \left\{ 4 \left(1 - \frac{A^2 E_0}{\rho_0} \right) \frac{\partial^2 u^1}{\partial \vartheta^2} - \frac{12\beta_0 A^2}{\rho_0} u^0 u^1 - A^2 \nu^2 \left[1 - b\frac{A^2 \mu_0}{\rho_0} \right] \frac{\partial^2 u^1}{\partial \vartheta^2} \right\} =$$

$$= \frac{\partial}{\partial \vartheta} \left\{ F^0 \frac{\partial(\rho A)}{\rho \partial X} + 2A \frac{\partial F^0}{\partial X} - 2\frac{bA^3 \nu^2 \beta_0}{\rho_0} \frac{\partial^3 u^0}{\partial X \partial \vartheta^2} - \frac{bA^2}{\rho} \frac{\partial(A\mu\nu^2)}{\partial X} \frac{\partial^2 u^0}{\partial \vartheta^2} \right\}.$$

We introduce new variable $\varphi \equiv \eta(\vartheta - \vartheta_0)$, then $\partial/\partial\vartheta = \eta(X)\partial/\partial\varphi$ and integrating the equation twice in φ we obtain the following linear inhomogeneous Legendre equation for perturbation $u^1(\varphi)$:

$$A^2 \nu^2 \left(1 - \frac{bA^2 \mu_0}{\rho_0} \right) \eta^2 u^1_{\varphi\varphi} - 4 \left[1 - \frac{A^2 E_0}{\rho_0} - \frac{3A^2 \beta_0}{\rho_0} \alpha \cosh^{-2} \varphi \right] u^1 =$$

$$C_1(1 - \tanh \varphi) + C_2 u^0 \varphi + C_3 u^0 \vartheta_{0,X} + b(C_4 \varphi u^0_{\varphi\varphi} + C_5 u^0_{\varphi} + C_6 u^0_{\varphi\varphi}),$$

where $u^0 = \alpha \cosh^{-2} \eta(\vartheta - \vartheta_0)$ and $C_i(X, E, \mu, \rho, \nu, \beta)$ are coefficients depending upon the slow variable.

To exclude the secular terms one should employ the condition (4.32) for orthogonality of solitary wave solution and the r.h.s. function $\Phi(u^0)$, that leads to the implicit ordinary differential equation to obtain the solitary wave amplitude $\alpha(X)$ in dependence on the inhomogeneity:

$$5\frac{\partial}{\partial X} \log \frac{\beta}{\alpha^3 \nu^2 (E + \alpha\beta)^2 (E + \alpha\beta - b\mu)} = b\frac{\alpha\beta\mu}{(E + \alpha\beta)(E + \alpha\beta - b\mu)}$$

$$\times \frac{\partial}{\partial X} \log \frac{\alpha^4 \mu^2 \nu^4}{(E + \alpha\beta)} \tag{4.37}$$

The solution $\alpha(X)$ is used to calculate the unknown function $A(X)$ defined by $A^2 = \rho/(E + \alpha\beta)$ and the wave length $\eta(X)$ according to the relationship $\eta^2 \nu^2 = \alpha\beta/(E - b\mu + \alpha\beta)$. The higher order approximation equations can be reduced to the linear inhomogeneous equations for associated Legendre functions and successfully solved. The solution $u = u^0 + \delta u^1$ vanishes for $z \to \infty$, but has the non-zero limit behind the soliton, i.e., the inhomogeneity leads to the shelf wave generation that has the amplitude of order δ and the velocity much smaller in comparison with the main pulse velocity.

Main results obtained by means of the asymptotic analysis of the wave propagation problem for an inhomogeneous rod are :

- there is the shelf behind the solitary wave occured due to the inhomogeneity. The harder the rod ($E'(X) > 0$ along the rod), the greater is the amplitude α of the compression soliton.

- the amplitude depends upon the variable nonlinearity parameter β, that was not found in several papers previously.

- a relationship exists between the values of *2nd* and *3d* order elastic moduli, that the amplitude α and the "energy" $I_2 = (1/2) \int u^2 dx$ of the initial solitary pulse both remain constant, while the sign of a shelf height follows to the sign of elastic moduli perturbation. Writing it for simplicity only for small values of amplitude α after corresponding power series expansion, we obtain

$$\beta^4 \nu^2 E^{-3} = const. \tag{4.38}$$

Moreover, when the inhomogeneity is that

$$\beta^4 E^{-3} = const \tag{4.39}$$

there will be *no shelf behind* an initial soliton, that spreads slowly, saving its bell shape and reducing its amplitude.

Further analysis will be done in Chapter 6 using numerical simulations.

4.4   Experiments in soliton propagation in the non-uniform rod

The solitary wave propagation in inhomogeneous rods may be of considerable interest for applications. For this reason several physical experiments were performed to confirm or reject some conclusions made in theoretical consideration. We have shown that the soliton propagates without change of shape in a uniform rod while its shape will vary in presence of inhomogeneities. In the last case the amplification or focusing may occur; in other words, the soliton amplitude will increase while its width will decrease. Then the localized area of plasticity and even fracture of a wave guide may appear, that can be of practical importance.

Successful experimental generation of a strain soliton in a rod with varying cross section was not mentioned even recently except in several papers by Dreiden et al. (1988, 1995). Nevertheless the strain soliton has been generated and observed in a uniform homogeneous nonlinearly elastic rod, using an experimental setup described in Chapter 4 above, and it is of interest to observe the strain soliton focusing in solids. Here we shall follow the explanation available in the paper (Samsonov et al. 1998).

For physical experiments the dimensional form of the model equation is more convenient, while the Lagrangian may be written without higher order nonlinear and differential terms in the relationships for kinetic energy K and potential strain energy Π. Substituting the Murnaghan model description into the formulae for K and Π, one can find respectively:

$$K = \frac{\rho_0}{2}\left(U_t^2 + \nu r^2\left[U_t U_{xxt} + \nu U_{xt}^2\right]\right), \tag{4.40}$$

$$\Pi = \frac{1}{2}\left(EU_x^2 + \frac{\beta}{3}U_x^3 + \nu E r^2 U_x U_{xxx}\right), \tag{4.41}$$

where $\beta = 3E + 2l(1-2\nu)^3 + 4m(1+\nu)^2(1-2\nu) + 6n\nu^2$ becomes the only coefficient depending on nonlinear elasticity of the rod. We have shown that the usage of truncated expansion (2.52) is sufficient to write relationships for K and Π. Substituting them into (3.32) and calculating $\delta S = 0$, one can obtain the following nonlinear equation in displacements:

$$U_{tt} - \frac{c_0^2}{R^2}\frac{\partial}{\partial x}\left[R^2 U_x\right] = \frac{1}{R^2}\frac{\partial}{\partial x}\left[\frac{\beta}{2\rho_0}R^2 U_x^2 - \frac{\nu}{4}\frac{\partial}{\partial x}(R^4 U_{tt}) + \frac{\nu^2}{2}R^4 U_{xtt}\right]$$
$$+ \frac{1}{R^2}\frac{\partial}{\partial x}\left[\frac{\nu c_0^2}{4}\left(R^4 U_{xxx} + \frac{\partial^2}{\partial x^2}\left(R^4 U_x\right)\right)\right] - \frac{\nu R^2}{4}U_{xxtt}, \tag{4.42}$$

where $c_0^2 = E/\rho_0$.

Therefore the additional linear dispersive terms appear in the equation above due to the terms u_2, w_3, resulted after the boundary conditions fulfillment at the free lateral surface, and we obtained the *refined and perturbed DDE*.

Let us consider now the rod which cross section varies slowly along the $x-$ axis, that will be described by a function $R = R(\gamma x)$, $\gamma \ll 1$. Introducing the notations: $u = U_x$, $\tau = tc_*$ and differentiating the equation (4.42) in x , we obtain a new equation

$$u_{\tau\tau} - \frac{\partial}{\partial x}\frac{1}{R^2}\frac{\partial}{\partial x}\left(R^2 u + \frac{\beta R^2}{2E}(u^2) + aR^4 u_{\tau\tau} - bR^4 u_{xx} - 4bR^3 R_x u_x\right) = 0, \tag{4.43}$$

with coefficients $a = -(\nu(1-\nu))/2$, $b = -\nu/2$, slightly deviating from those obtained in (Samsonov, 1988): $a = \nu^2/2$, $b = \nu^2/(2(1+\nu))$. Two first terms here describe the common linear wave, the third governs the nonlinearity, the two following terms are responsible for dispersive features of a wave guide, while the last term, being of the same order, looks like the dissipative one, but occurs due to the cross section variation.

The uniformly valid asymptotic analysis performed in the previous subsection for the 4th order perturbed p.d.e. can be used directly to solve (4.43). To describe the evolution of a travelling strain wave u we introduce the phase variable θ and the slow variable $X \equiv \gamma x$, as follows:

$$\theta_\tau = -1, \quad \theta_x = A(X). \tag{4.44}$$

The solution to equation (4.43) will be found in new variables in the power series in γ:

$$u = u_0 + \gamma u_1 + \ldots . \tag{4.45}$$

Substitution of it into the equation (4.43), gives in the leading order of γ the well known dimensional solitary wave solution for u_0:

$$u_0 = \frac{3E}{\beta} \alpha \, \cosh^{-2} \left(k(X)[\theta - \theta_0(X)] \right) ,, \tag{4.46}$$

depending upon the unknown function $\alpha = \alpha(X)$, $\alpha > 0$, while both A, k are expressed through it by means of the formuli:

$$A^2 = \frac{1}{1+\alpha}, \quad k^2 = \frac{\alpha(1+\alpha)}{4R^2[a(1+\alpha) - b]}. \tag{4.47}$$

Both A and k should be and will be real in (3.49) for most of conventional elastic materials (with $\nu > 0$), if:

$$0 < \alpha < \frac{\nu}{1-\nu}, \tag{4.48}$$

As it was shown before, the type of the strain wave (4.46) (compressive or tensile one) is defined by sgnβ, which depends on the elasticity of the rod material, respectively.

Let us study a distortion of the solitary strain wave due to the non-uniformity or 'geometrical' inhomogeneity considered. The following differential equation for α arises from the secular terms absence condition in order $O(\gamma)$:

$$\frac{\partial}{\partial X} \ln \frac{R^2 \alpha^2}{2kA^3} + \frac{4bk^2 R^2 A^4}{5} \frac{\partial}{\partial X} \ln 2R^4 \alpha^2 Ak = 0, \tag{4.49}$$

that after use of (4.47) is reduced to a nonlinear 1st order o.d.e. for the amplitude variation

$$\frac{R_X}{R} = \alpha_X \left(\frac{1}{6(1 - D + \alpha)} - \frac{1}{2\alpha} - \frac{1}{3(1 - D_1 + \alpha)} - \frac{1}{3(1 - D_2 + \alpha)} \right), \tag{4.50}$$

where:

$$D = \frac{b}{a} = \frac{1}{(1-\nu)}, \quad D_{1,2} = \frac{2 \pm \sqrt{9 - 5\nu}}{5(1-\nu)}.$$

Taking the restrictions for α (4.48) into account, we conclude that the expansion in brackets in the right hand side of (4.50) is always positive. Therefore the *magnitude* of the soliton will increase with the radius decrease, as we have seen, dealing with the formal asymptotic analysis. The equation (4.50) can be integrated directly, that yields

$$\frac{R^6 \alpha^3 \left[\nu + \alpha(2\nu - 6/5) - \alpha^2(1 - \nu) \right]^2}{(1-\nu)[\nu - \alpha(1 - \nu)]} = \text{const.} \tag{4.51}$$

The routine analysis of the functions u_0 (4.46) and $\partial u_0/\partial x$ shows that the distortion of the wave shape takes place apart from the amplitude variation. When the bell-shaped soliton propagates along the tapered rod, its front side becomes steeper while the back one becomes smoother. Vice versa, the front side of the solitary wave, moving along the expanding rod, becomes smoother, while the back one becomes steeper. The equation for determination of an extremum of a derivative $\partial u_0/\partial x$

$$\gamma\frac{R_X}{R} + [k(1 - \gamma\theta_{0,X}) + \gamma k_X[\theta - \theta_0(X)]] \tanh(k[\theta - \theta_0(X)]) = 0. \qquad (4.52)$$

shows that for the wave propagation along the narrowing rod ($R_X < 0$) the extremum is achieved for $\theta - \theta_0(X) > 0$, while along the extending rod ($R_X > 0$) - for an inverse sign of that expression. Then the soliton accelerates in narrowing rod and decelerates in the expanding one in comparison with the same soliton moving along a uniform (homogeneous) rod.

Exact formulae (4.47), (4.51) may be easily simplified to analyse the wave parameters variations detectable in experiments. We have seen that the strain wave amplitude has to be restricted with a physical condition of the strain's elasticity:

$$|\sqrt{1 + 2C_{xx}} - 1| < e_0, \qquad (4.53)$$

where e_0 is the yield point of a material and lies in the interval $10^{-4} - 10^{-3}$ (Frantsevich et al. 1982). Therefore α will have to be small enough to lead to the following approximations from (4.47), (4.51):

$$A = 1, \quad k^2 = \frac{\alpha}{4R^2(a - b)}, \quad \frac{\alpha}{\alpha_0} = \left(\frac{R_0}{R}\right)^2. \qquad (4.54)$$

The most important feature of the next order asymptotic solution is in the appearance of a plateau, propagating behind the soliton (3.48) with much less velocity. The difference in the values of a, b calculated in the framework of two theories results in the quantitative deviation in the plateau amplitude value. However, it is of order $O(\gamma)$, hence its changes will be small also in comparison with the value obtained in (Samsonov, 1997). This deviation seems unlikely to be detected by means of the experimental arrangements used, and this part of solution is omitted here.

To perform the experiments in soliton propagation along the tapered elastic rod we have used the same experimental arrangements as described in Section 4.2.

The holographic wave pattern recorded in experiments contain horisontal black areas above and below a central fringe area which are due to cylindrical 'caps' along the rod, see Figure 4.16.

Black rectangular frames (as well as the grey frames in Figures 4.17, 4.18 below) surrounded the fringe pattern inside a rod appear due to the fact that lateral surface of a rod beyond the central area of observation is not precisely perpendicular to the laser beam, i.e., not transparent and therefore reflects the light.

A choice of the rod's cross sections variation is caused by two factors. Firstly, we were going to observe a geometrical inhomogeneity influence just on the strain soliton, secondly, the experimental setup limitations should be taken into account.

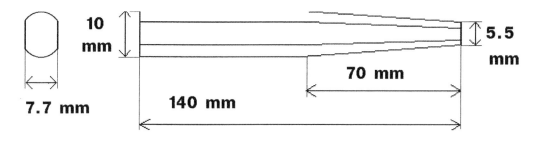

Figure 4.16: Geometry of the tapered rod

Measurements of the soliton amplitude in a homogeneous rod resulted in the estimation of the parameter $\varepsilon = O(10^{-3})$. When the inhomogeneity parameter γ is chosen to be $\gamma << \varepsilon$, then the possible variation of the initial rod radius ($R_0 = 5$ mm) at the distance 100 mm along the axis will be of order 0.1 mm or 2% from the initial value. The estimation of the amplitude change in this case by means of an approximation (4.54) shows that such magnitude corresponds to the oscillations of the observed solitary wave front (Dreiden et al., 1995). So it seems hardly to be possible to detect such a deviation using our experimental setup. Therefore we had to choose the inhomogeneity parameter as follows $\gamma >> \varepsilon$.

In addition, we note that the nonstationary process takes place in experiments in contrast to quasistationary process governed by the asymptotic solution obtained above. When $\gamma >> \varepsilon$ the inhomogeneity will change the initial pulse shape earlier than both nonlinearity and dispersion will, and the strain soliton will hardly appear from an initial weak shock. Thus the rod cross section should remain constant at the distance required for the soliton generation and separation, and begin vary only after it. Experiments on the soliton generation in a homogeneous rod show that soliton appears even at the distance of 60 mm ($12 R_0$) approximately from the input edge of the rod.

Based on this analysis, the polystyrene rod of 140 mm long was made with the uniform and the narrowing parts, as it is shown in Figure 4.16, and two cut-offs were made on the lateral surface for the observation purposes. The rod radius decreases linearly from the value $R_0 = 5$ mm to the value $R = 2.75$ mm along the distance 70 mm. In this case the inhomogeneity parameter $\gamma = 0.032$ is much greater than the typical soliton amplitude $\approx 10^{-4}$ for the homogeneous rod.

We consider the holographic interferograms of the longitudinal strain soliton recorded in the transition interval from the rod with uniform cross section to its tapered part at the distance $40 - 90$ mm from the edge of the rod (Figure 4.17), and in the interval, $75 - 125$ mm (Figure 4.18), where the rod is tapered. Diameter of the recording beam is equal to 50 mm as before. For convenient experimental data processing, one of the disturbed interference fringes inside the rod (marked with arrows) was extracted from and placed below the interferogram. Fringes in a surrounding liquid remain undisturbed (horizontal), that confirms that the observed wave propagates inside the rod. The shape of the strain wave was reconstructed by means of (4.8) using the following values of parameters: $n_0 = 1.33$, $n_1 = 1.6$, $\Lambda = 6.9 \cdot 10^{-7}$ m,

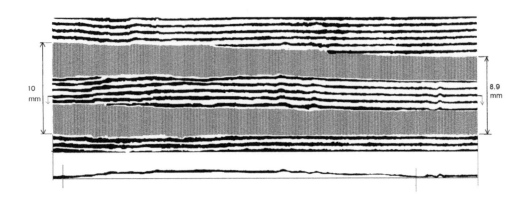

Figure 4.17: Interferogram of the strain soliton in the nonlinearly elastic rod recorded at the beginning of the tapered part.

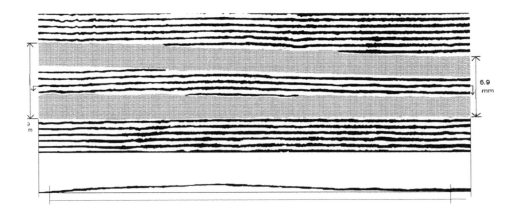

Figure 4.18: Interferogram of the strain soliton recorded in the tapered nonlinearly elastic rod.

$\nu = 0.35$. It must be taken into account that light passes the *different* distances $2h$ in different cross sections. At the interval where the cross section remains uniform, we have $2h = 2h_0 = 7.75 \cdot 10^{-3}$ m, while the measured cross sections for the tapered rod's part are shown in Table 4.3.

Table 4.3. Light path distance measurements data inside the tapered rod.

Distance from the entire edge, mm	Diameter of the rod, mm	Light path $2h$, mm
50	10	7.75
65	10	7.75
70	10	7.75
75	9.8	7.45
80	9.5	7.2
85	9.2	7.0
90	8.9	6.8
95	8.6	6.6
100	8.3	6.3
105	8.0	6.1
110	7.7	5.8
115	7.5	5.7
120	7.1	5.4
125	6.9	5.0

One can see that the maximal fringe shift on both interferograms is almost equal to the width between two neighbouring fringes, i.e., to one fringe width. Substituting the data from Table 4.3 into (4.8), where $2h$ varies now from one cross section to another, one can calculate finally the soliton parameters and obtain the soliton evolution in the tapered rod, see Figure 4.19. The envelope lines are drawn there after interpolation. For convenience, the compressive waves, having the negative amplitude, are shown in the first quadrant.

Thus, using the laser generator of weak shock waves and the holographic setup, we have made probably the first generation, detection and records of the focusing of strain solitary wave inside the nonlinearly elastic tapered rod.

The following arguments may confirm the observation of the genuine strain solitary wave in our experiments. As before, there is no tensile area behind the observed long compressive wave (having a length $\lambda > 7\,R$), that is a typical feature of the localized nonlinear wave. Tensile areas, if any, can be easily detected using the same apparatus: the fringes will be shifted in an opposite direction. However, deformation of the rod behind the soliton was studied in detail, and nothing was observed there except straight fringes, i.e., the rod was free of strain again after the soliton propagation.

Secondly, even at the distances, exceeding dozens of rod's radii, both the shape and the wave parameters remain permanent and do not exhibit any essential distortions in the uniform rod, that is the nonlinear strain wave possesses one of the most distinctive feature of a soliton. The distance chosen for our observations seems to be sufficiently large because polystyrene is well known to be an effective absorber of acoustic and

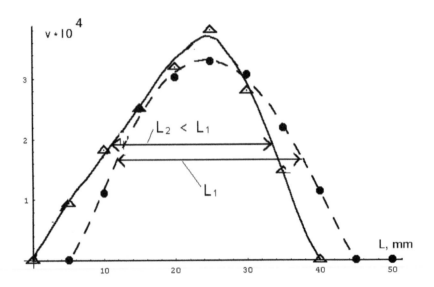

Figure 4.19: Focusing of longitudinal strain soliton. After experiments two graphs of strain vs. soliton width are drawn after interpolation. Solid circles and the dashed interpolative line both correspond to data measured on a 40 − 90 mm interval of the rod's length. Open triangles and the solid interpolative line correspond to them on a 75 − 125 mm interval.

shock waves. The last was confirmed by considerable decay of the shock wave, which moved ahead of the soliton, as shown in Figures 4.17, 4.18.

The enlargement of the amplitude scale allows us to visualize main features of the solitary wave in the tapered rod, see Figure 4.19. All the features predicted by our theory appear in experiments, namely, the increase of the amplitude, the steepness of the wave front and smoothness of its back, i.e., asymmetric deformation of the bell-shaped soliton. Moreover, the characteristic width $L_1 = 25, 2$ mm of the pulse shown in Figure 4.17 in the homogeneous part of the rod at the one-half amplitude level is visibly greater than the similar value, $L_2 = 22, 3$ mm, in the narrowing part, hence the width of the localized strain solitary pulse *decreases* along the tapered rod.

Eventually, simultaneous increase of the amplitude and decrease of the width (i.e., the focusing) are distinctive features of the soliton in tapered wave guide, while the parameters of the linear strain wave are independent and defined by the initial or boundary conditions only.

The abilities of our experimental setup do not allow to measure directly the soliton acceleration caused by the narrowing cross section along the rod.

However all other details of the distortion of the wave observed (Figure 4.19) compared with those theoretically predicted for the strain soliton (3.48) lead to the conclusion that both the strain soliton and its focusing were observed, indeed, in our experiments. Fortunately we detected also both steepening of the soliton front and simultaneous smoothening of its back in close correspondance with theoretical predicition of a soliton shape variation when focusing.

Therefore, we have shown the possibility *to transfer and to concentrate the strain energy* at long distances without losses even in materials, having a considerable absorption (dissipation) for shocks. Presumably it means that the nonlinear absorbtion is much less than the linear one that does not affect the soliton, moving with transonic velocity; this problem requires further analysis.

The measurement of the wave amplitude is supposed to be quite plausible for the comparison with the theory. One can see in Figures 4.17, 4.18 that the maximal amplitude of the strain soliton is achieved at the distances 60 and 95 mm from the rod input edge, respectively. Then from the estimation of the amplitude we obtain it equal to $3.29 \cdot 10^{-4}$ in the interval 40-90 mm, and to $3.83 \cdot 10^{-4}$ for the interval 75-125 mm. Therefore, the soliton magnitude increases 1.16 times. The estimation using the simplified formulae (4.54), and the length dependence of the kind $R = R_0 - \gamma(x - 70)$ gives the amplification 1.31 times, which is in good agreement with the experimental data.

However, some new theoretical results cannot be checked in our experiments and require further study, namely the observation of the amplitude dependence upon the Poisson coefficient ν, is expected to be of interest for applications found for the first time by means of the exact formula (3.52).

The advantages of the theoretical description proposed here are of importance for the study of the periodical, particularly, cnoidal waves, because the deviation in the values of wave lengths can be measured in experiments with reasonable accuracy. Another problem for which the theory should be applied is the wave propagation along a wave guide embedded in an external medium. The theory also may be used for the soliton focusing study in 2D wave guides, in particular, in plates and layers.

Chapter 5

Nonlinear waves in complex wave guides

This Chapter will focus on a discussion of the nonlinear wave propagation problems which are not governed by the 1+1D nonlinear quasi hyperbolic equation, but by more complicated statements. The wave problem in 1+2D wave guide, i.e., the longitudinal strain wave propagation in a nonlinearly elastic thin plate will be studied and brought to the explanation of the first experimental observation of the long 2D solitons. The long nonlinear waves in a thin layer superimposed on the elastic half space and in a rod embedded into an elastic surrounding medium will be studied formally and subject to the numerical simulation. All these problems are of interest for applications, however it will be shown they require much more complicated analytical treatment, and the combination with numerical modelling is very helpful.

5.1   Longitudinal nonlinear waves in an elastic plate

Now we shall consider the 2-dimensional problem of the solitary wave generation and observation. To do it we shall solve the nonlinear coupled equations governing the longitudinal strain wave propagation in a nonlinearly elastic thin plate made of Murnaghan (compressible) or Mooney (incompressible) material. We shall not use neither any assumptions about small V in comparison with U nor asymptotic limit $\varepsilon \to 0$, instead, we shall solve explicitly the equations with periodic or zero boundary conditions for strains at infinity calculated for periodic or localized solutions respectively.

Starting with equations (2.62) and (2.68) derived in Chapter 2, we introduce the following short notations for the strains:

$$u \equiv U_x, \ v \equiv V_y, \ p \equiv u_y, \ q \equiv v_x$$

and the phase variable $\theta = x \pm Vt \pm ky$, in which expression both the velocity V (not a displacement here !) and the arbitrary angle k defining the wave propagation direction are assumed to be constant.

Then the equation (2.62) governing the nonlinear travelling long waves in the Mooney plate yields:

$$\left[\left(V^2 - 1 - k^2/4\right)u - \frac{3v}{4} - \epsilon\left(\frac{4V^2 - 1 - k^2}{48}s_{\theta\theta} + P \pm k^2 Q\right)\right]_{\theta\theta} = 0, \qquad (5.1)$$

$$\left[\left(V^2 - 1 - k^2/4\right)v - \frac{3uk^2}{4} - \epsilon\left(k^2\frac{4V^2 - 1 - k^2}{48}s_{\theta\theta} + k^2 P_1 \pm k^2 Q_1\right)\right]_{\theta\theta} = 0, \qquad (5.2)$$

and in addition:

$$q = \pm\frac{v}{k}; \ p = \pm ku \qquad (5.3)$$

where expressions for P_1 and Q_1 can be obtained from P and Q respectively by means of substitutions described after (2.62) in Chapter 2.

We shall show that the system (5.1), (5.2) can be reduced to a single nonlinear equation with respect to $u(\theta)$, see (Samsonov, 1993a; Sokurinskaya, 1994). Indeed, subtracting the second equation from the first one multiplied by k^2, the result thus obtained will lead, after two integrations with the boundary conditions being taken into account, to two possible simple relationships between strains $u(\theta)$ and $v(\theta)$, namely:

$$v = uk^2, \qquad (5.4)$$

or

$$(k^2 + 1)^2 - 4V^2 = \varepsilon(1 - \beta)(u + v)(k^2 + 1)^2$$

where β is the only 3d order modulus. Note that k determines also the ratio of strains u and v and may be arbitrary, not small.

It is easy to show by substitution that the second relation reduces the problem to a linear one and is of minor interest for further discussion. Taking the *first* relationship (5.4) into account explicitly enables us to reduce the system (2.62) in the travelling wave case to a single equation with respect to the strain $u = U_x$:

$$u_{\theta\theta} + au^2 + 12bu = 0, \qquad (5.5)$$

where β is the 3d order modulus,

$$a = 72\alpha^2/(\alpha^2 - 4V^2), \quad b = 4(V^2 - \alpha^2)[\varepsilon\alpha^2(\alpha^2 - 4V^2)]^{-1}; \quad \alpha^2 = 1 + k^2.$$

Obviously, by substitution $W(\theta) = -b - ua/6$ the equation (5.5) is reduced to the Weierstrass equation:

$$W''_{\theta\theta} = 6W^2 - 6b^2 \qquad (5.6)$$

and can be solved in terms of the Weierstrass elliptic function $\wp(\theta + const; g_2; g_3)$. These solutions to (5.6) have been investigated in details in Chapter 3, and the existence has been demonstrated of both one-parameter and two-parameter solutions of the type of localized or periodic discontinuities, of the cnoidal, harmonic and solitary waves. All of them are, in fact, the particular cases of the solution in terms of the Weierstrass function $\wp$.

Therefore we used the intrinsic symmetry of the 2D travelling wave propagation problem in the plate and reduced it to the only Weierstrass equation.

Hence, the expression for the solitary wave, as a particular form of the Weierstrass function solution, in an incompressible Mooney plate after changes of variables is written as follows:

$$u = A \cosh^{-2}\left[(\theta + \text{const})\sqrt{\frac{12A}{4\varepsilon\alpha^2 A - 3}}\right];$$

$$v = uk^2, \tag{5.7}$$

where the solitary wave amplitude may be written as $A = 1 - V^2/(\varepsilon\alpha^2)$ that gives the necessity to re-consider the solution in terms of velocity, if necessary. The condition for localization of the wave (5.7) results in the following restrictions for velocity:

$$V^2 > 1, \text{ or } 0 < V^2 < 1/4 \tag{5.8}$$

As before, we found the interval of velocities, with any of which the soliton cannot propagate, and the two-parameter solution (5.7), describes both the supersonic compression wave with $A < 0$ and $V^2 > 1$, and also the subsonic tensile wave with

$$0 < \frac{3}{4\varepsilon\alpha^2} < A < \frac{1}{\varepsilon\alpha^2} \text{ and } V^2 < 1/4$$

Solitary waves (5.7) with velocities from the 'dead zone' $1/4 \le V^2 \le 1$ cannot propagate, while in this case the other strain soliton solution exists in the form

$$u = \frac{2A}{3} - A \cosh^{-2}\left[(\theta + \text{const})\sqrt{\frac{12A}{3 - 4\epsilon\alpha^2 A}}\right] \tag{5.9}$$

i.e., the soliton on the pedestal, that provides a constant deformation for $|\theta| \to \infty$.

It was quite surprising that the reduction of the coupled nonlinear wave equations and derivation of solution described above is valid also for materials governed by the Murnaghan model, see (2.68), with the only difference in the coefficients in (5.5), which should be substituted now with the following values:

$$a = \frac{36\alpha^2(1 + \beta_2)(1 - \nu)^2}{\nu^2[2V^2 - (1 - \nu)\alpha^2]}, \quad b = -\frac{2(V^2 - \alpha^2)(1 - \nu)^2}{\varepsilon\alpha^2\nu^2[2V^2 - (1 - \nu)\alpha^2]}; \quad \alpha^2 \equiv 1 + k^2.$$

Therefore, in the compressible (Murnaghan) nonlinearly elastic plate, the general wave solution in terms of the Weierstrass function and, in particular, the solitary strain wave can be found in a way similar to that used above. A two-dimensional strain soliton in the compressible Murnaghan plate has the form

$$u = \frac{A}{1 + \beta_2}\cosh^{-2}\left[(\theta + \text{const})\frac{1 - \nu}{\nu}\sqrt{\frac{6A}{1 + \nu + 2\varepsilon\alpha^2 A}}\right] \tag{5.10}$$

where the amplitude is now equal to

$$A = \frac{V^2 - 1}{\varepsilon\alpha^2}$$

and exists if the wave velocity V is subjected to the restrictions

$$V^2 > 1, \text{ or } 0 < V^2 < \frac{1-\nu}{2},$$

that for the limiting value $\nu = 0.5$ and $\beta_2 = -2$ coincides with the solution (5.7) for incompressible Mooney plates. For $\beta_2 > -1$ (as it is for PMMA, melted quartz, glass) the solution (5.10) describes the propagation of either *supersonic tensile* pulse with $V^2 > 1$ and $A > 0$ or *subsonic compression* wave with $A < 0$, $V^2 < (1-\nu)/2$. For many metals $\beta_2 < -1$ and the solution (5.10), as well as the solution (5.7) for incompressible plates, describes transonic solitary compression wave ($A > 0$) or subsonic tensile waves ($A < 0$). The increment in the velocity of the nonlinear wave in the plate compared with that in a linear problem is of order $O(\varepsilon)$.

It may be emphasized that the parameters of the exact localized solution (5.7) or (5.9) of the system of nonlinear equations differ significantly from an approximate solution obtained as a result of reduction of the initial system to the Kadomtsev-Petviashvili equation. It can be shown that the differences between the exact and approximate values of the amplitude and the localized pulse width are of the same order $O(\varepsilon)$ as the differences arising due to the refinement of the elasticity problem from the linear statement to the nonlinear one, and even in this case - only for velocities close to the velocity of the waves in a linearly elastic plate. Moreover, the condition of localization of the exact solution (5.7) defines two intervals of allowed velocities of pulses, while the similar condition for the solution of the KP equation leads to a single inequality of the form $V^2 > [\alpha^2 + \sqrt{2\alpha^2 - 1}]/2$ or, in an equivalent form, $A < 0$, i.e., the approximate solution does not describe tentative tensile waves at all. Therefore the common description (Potapov and Soldatov, 1984) of the two-dimensional nonlinear deformation waves based on the KP equation seems to be insufficient.

Note that the deflection displacement is calculated using the parameters of the longitudinal wave propagating along the midplane of the plate:

$$W = \varepsilon U = \varepsilon V$$

i.e., has the solitary wave features, however is of order $O(\varepsilon)$ with respect to the displacemets in the middle plane of the plate. Another important thing to be noted is that k is not fixed in the solutions and may be chosen as $k = 1$. It means that by rotating the initial coordinate system, one can get a wave propagating along an arbitrary line in the middle plane of the plate, e.g., along a co-ordinate axis.

5.1.1   Generation and observation of the strain soliton in a plate

Having the estimations derived above, the physical experiments in generation and observation of the solitary strain wave in the 2D wave guide, e.g., in the Murnaghan plate, will be considered.

As we have mentioned before, it is particularly interesting to study the fast intense loading of polymers, such as phenylone and polystyrene, used in the aerospace industry, thermonuclear layered target manufacturing, etc. due to their wear and radiation resistance. Of main interest are the nonlinear waves propagating along a 2D wave guide without distortion, which can be dangerous due to elastic energy transfer over large distances practically without losses, even in materials having strong linear dissipative features.

At first glance, after experiments with solitary wave generation in a circular straight cylinder, the success of experiments in excitation of similar waves in a plate is almost guaranteed due to similarity in theoretical description of longitudinal strain waves in a rod and of plane longitudinal waves in a plate.

However, we need to keep in mind that the excitation of the plane wave requires a short powerful and non-destructive initial pulse having the front parallel to the entire end of the plate. This problem may be solved easily using the metallized film target and a water gap between the film and the entire end, as it was done previously in the rod problem. The interferogram of the wave pattern near the target in water (see Figure 4.5) allows the conclusion that the wave front of initial weak shock has a long planar part that will be used to load the input cross section of a plate. This part is much larger than the diameter of the rod used in previous experiments, while smaller than maximal linear size of the end cross section of the plate. However, the entire region is covered by an observation window (50 mm in diameter), through which we can see a fairly plane nonlinear longitudinal strain wave near the central part of the plate. The front distortion becomes recognizable over distances much greater than 50 mm. In our experiments we used a 120 $\times$ 120 $\times$ 11 mm polystyrene plate so that the shock acted on its 120 $\times$ 11 end. Polystyrene behaviour is described satisfactory in the frame of Murnaghan's model.

We position the co-ordinate system as it was shown in Figure 2.1. Let a weak shock be propagated along the x axis in a nonlinear elastic plate, resulting in formation of a plane localized strain wave, a strain soliton u_x. We need to obtain a relationship between the holographic fringe shift and the amplitude of a soliton in order to interprete the experimental data.

Using the formula (2.71) for the relation between transversal displacements and the in-plane strains:

$$W = -\frac{\nu z}{1 - \nu}(u + v),$$

rewriting it in terms of the phase variable $\theta \equiv x \pm ky \pm Vt$, where V- the travelling wave velocity, t is time, we get

$$u + v = u(1 + k^2). \tag{5.11}$$

The analysis given above leads to the solitary wave solution in the form (5.10), and we should derive a formula linking the fringe shift ΔK resulted from the light phase difference λ in deformed $\Delta\phi_2$ and in reference states $\Delta\phi_1$:

$$\Delta K = \frac{\Delta\phi_2 - \Delta\phi_1}{2\pi} \tag{5.12}$$

For polystyrene and most metals $\beta < -1$ therefore the solution describes the transonic solitary compression waves (with $A > 0$), i.e., the longitudinal density waves.

The apparatus described above allows the determination of the strains in a sample from the intereference pattern. The phase shifts are caused by the light propagation through media with different refraction indices, namely, water with n_0, undeformed plate with n_1 or stressed plate with n_2. As a result we get:

$$
\begin{aligned}
\Delta\phi_1 &= \frac{2\pi}{\lambda}((L-h)n_0 + hn_1), \\
\Delta\phi_2 &= \frac{2\pi}{\lambda}((L-h-2\Delta h)n_0 + (h+2\Delta h)n_2),
\end{aligned}
\tag{5.13}
$$

where it was taken into account that the soliton leads to the local increase of plate's thickness in $2\Delta h$, where L is the cell thickness. The undetermined refractive index of stressed material n_2 is defined from the relative change in the density of the plate $\delta\rho$:

$$
n_2 - n_1 = -(n_1 - 1)\frac{\rho_0 - \rho}{\rho} \equiv -(n_1 - 1)\delta\rho.
\tag{5.14}
$$

The masses of elementary volumes before and after deformation remain constant, then the volume expansion D is equal to relative variation of density $\delta\rho$ caused by the wave propagation along the x axis as $D = \delta\rho$. On the other hand, D is also the ratio of the elementary volumes difference $dV - dv$ in the current and reference configurations to the volume dv in the reference frame:

$$
D = (1 + 2I_1 + 4I_2 + 8I_3)^{1/2} - 1,
\tag{5.15}
$$

where I_k are invariants $I_k(\mathbf{C})$ of the finite elasticity tensor $\mathbf{C}$.

Since Δh is identical to the displacement W on the plate surface, then using (5.11),(5.12), (5.14), we obtain the following exact relation between the fringe shift and strain ΔK :

$$
\Delta K = \frac{\nu h(u+v)}{1-\nu}(n_1 - n_0 - D(n_1 - 1)) - hD(n_1 - 1).
\tag{5.16}
$$

Strains are assumed to be relatively small, then it can be simplified neglecting terms of order $O(u_x^2)$, that leads to the explicit formula:

$$
A = \max u = -\lambda\Delta K \left\{ \frac{h(1+k^2)}{1-\nu}[(1-2\nu)(n_1 - 1) + \nu(n_1 - n_0)] \right\}^{-1}.
\tag{5.17}
$$

which is similar to the amplitude estimation given above for experiments in Murnaghan's rod.

In our case $k = 0$, and the strain solitary wave amplitude corresponds to the maximal value of the right hand side expression. The solitary wave length is measured directly in the interferogram as it was in the rod problem.

As before, the power density of the laser beam acting on the target was defined by the energy meter, kept constant throughout the experiment and equal to $2.3 \cdot 10^8 W/cm^2$. The wave propagation process was recorded in the holograms in the direction perpendicular to the wave using the OGM-20 Q-switched pulse ruby laser.

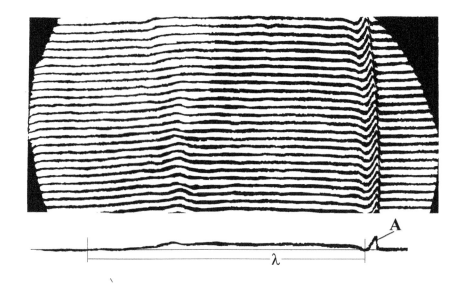

Figure 5.1: Interferogram of longitudinal plane strain wave in the plate. The smooth fringe shift represents the solitary strain wave, propagating to the right, and is shown also in the footnote beneath the photograph. Left edge of the photograph is in 40 mm from the entry face of the plate, the right one in 90 mm.

In Figures 5.1, 5.2 one can see the holographic interferograms of a soliton at the following distances from the entry face: 40-90 mm and 75- 125 mm. The smooth and distant fringe shift (a soliton, or rigorously, the solitary wave of strain) is selected and shown in the footnote beneath each hologram. The figures show a primary shock wave A, that initiated the compression soliton, and also a secondary shock of complex profile added to the solitary wave in the plate, which looks like a small elevation 'on horseback' in the centre of the soliton. It is the result of the primary shock in the liquid being reflected from the end of the plate, then from the target and later entering the plate also. It may be easily proved via calculations of time necessary for a shock to go this distance.

These holograms allow for study of the evolution of the wave pattern in the plate and to observe the soliton formation during the wave propagation along the 2D wave guide.

The shock A attenuates, propagating in the plate, while its velocity remains bigger than the velocities of both linear sound wave and the soliton - the soliton lags behing from one frame to another. The solitary wave itself is very long longitudinal through-shaped wave of compression, not followed by any tensile wave. Its shape and length are similar to those of a soliton in a rod. Calculations of the parameters of the soliton (Figure 5.1) yield the values of the amplitude : $A = -1,15 \cdot 10^{-4}$, and the length $\Lambda = 32,6$ mm, and the treatment of data obtained at different distances reveals almost the same values, that confirms the permanent shape feature, one distinctive feature of the solitary strain wave.

It is to be noted that the absence of any transversal perturbations of fringes confirms the assumption about the possibility of generation of the plane solitary wave,

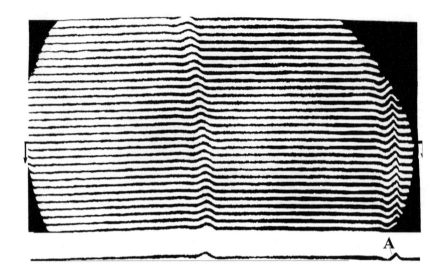

Figure 5.2: The same as in previous figure. Left edge of the picture is in 75 mm from the entry face of the plate, the right one in 125 mm. The soliton comes off the dispersing shock wave A. Soliton shape, amplitude and length remain permanent, the lag increases.

at least, in the central part of the plate, made above. We studied carefully the strain field behind the soliton - the fringes remain horizontal, that means that there is no wave of opposite sign behind the soliton.

The soliton amplitude in the plate is 2.45 times *less than in the rod* made of the same polystyrene and under identical excitation conditions. The strain soliton in the rod (see Section above) has an amplitude $A = -3.2 \cdot 10^{-4}$. It allows the suggestion that some energy of initial pulse propagates along the entry of the plate due to total internal reflection as well as inside the plate perpendicular to the wave propagation direction, i.e., subjects to diffraction.

Possible inhomogeneity of the 2D wave guide can influence the solitary wave propagation strongly, and it may be studied in a way similar to the asymptotic analysis made in the problem of inhomogeneous rod, or, using the numerical simulation approach.

5.2 Longitudinal waves in rods embedded in a surrounding medium

We derived in Chapter 2 and studied in Chapter 3 the exact solutions to the DDE with dissipative terms, or, generally, to nonlinear quasi hyperbolic dissipative equations and to some nonlinear parabolic equations. Hyperbolic models are widely used for describing wave problems in active or dissipative media and provide the solutions as a discontinuous elliptic function of general type, a smooth kink, or a periodical wave, that require the physical interpretation.

We shall show that the statement and the solutions to the wave dynamics problem in a wave guide embedded in a surrounding medium depend essentially on the type of contact between them.

5.2.1 Kinks in a rod in full contact with a surrounding medium

To begin with, we assume the lateral surface of the rod is in contact with a dissipative or active external medium, governed by the Kerr model, see Section 2.5, and the dissipation (of any sign) is sufficiently strong and may influence the balance between nonlinearity and dispersion.

As a result, we obtained an equation of second order in time governing nonlinear long waves. We present it here only in a nondimensional form:

$$4(w_{tt} - w_{xx}) = \varepsilon(6w^2 + aw_{tt} - bw_{xx} + dw_t)_{xx} \tag{5.18}$$

where w is proportional to strain $\partial U/\partial x$, ε is a small parameter, expressing the balance of nonlinear, dispersive and energy influx terms in a system; coefficients a, b are defined by the elasticity moduli, whereas the effective values of the dissipation parameter d and the shear modulus μ are $d = 0$, $\mu = \mu_0 + 4k_2/R$ for a passive external medium and $d = 4\eta X/ \left(R\sqrt{E\rho}\right) < 0$, $\mu = \mu_0$ for an active one. Equation (5.18) for $d = 0$ and for $d > 0$ was investigated and some exact solutions in the form of travelling waves were obtained, for physics we note that when the interaction of a rod and the elastic external medium ($d = 0$) is considered, the linear longitudinal velocity $c = \sqrt{E/\rho}$ and the shear wave velocity $c_1 = \sqrt{\mu/\rho}$ increase in comparison with the values corresponding to the free wave guide problem.

If the medium is active ($d < 0$), then the energy from it could be transferred into the energy of residual strains at wave fronts. Any positive value of d corresponds to the viscous external medium, and the energy of preliminary strain u_x^0 may compensate the energy absorption due to viscosity.

Let us consider some exact travelling wave solutions, depending upon the phase variable $z = x \pm Vt$ only, of equation (5.18) in the form

$$4(w_{tt} - w_{xx}) = (6w^2 + aw_{tt} - bw_{xx} + dw_t)_{xx} \tag{5.19}$$

as well as the general cubic DDE (cDDE) with the same energy influx:

$$4(w_{tt} - w_{xx}) = (Cw^3 + aw_{tt} - bw_{xx} + dw_t)_{xx} \tag{5.20}$$

under the following boundary conditions:

$$z \to -\infty \Rightarrow w \to w_1$$

$$z \to +\infty \Rightarrow w \to w_2$$

$$|z| \to \infty \Rightarrow \partial^k w/\partial z^k \to 0$$

The small parameter ε was excluded from (5.18) by means of scaling: $\varepsilon w \to w$, $\varepsilon a \to a$ etc, because we are not going to use any asymptotic expansions. After integrating equations (5.19), (5.20) twice with respect to z, we obtain the nonlinear ordinary differential equation with coefficient B before the first derivative, depending on d :

$$w'' = Bw' + Gw^3 + Dw^2 + Ew + F, \ B = -d(\pm V)/(aV^2 - b) \qquad (5.21)$$

We should note that (5.21) cannot be reduced to any equation of the Painlevé type, and two cases of equation (5.21) which can be integrated are $W'' = 2W^3$ for $G \neq 0$ and $W'' = 6W^2$ for $G = 0$.

Thus it was shown in Chapter 3 that the exact continuous travelling wave solution to (5.21) for $G = 0$ has the form:

$$w = w_1 + (w_2 - w_1)y^2(K + y)^{-2}, \ K > 0 - const, \ y \equiv \phi(z) - \exp(Bz/5). \qquad (5.22)$$

The height of the smooth jump (an algebraic kink) is $(w_2 - w_1) = 6B^2/(25D)$, and the velocity of the wave is $V^2 = 1 - 3E/(2D)$.

As for $G \neq 0$, the cubic DDE for active medium has the following exact solution:

$$w = w_1 + (w_2 - w_1)y/(K + y), \ K>0 - const, \ y = exp(Bz/3),$$
$$w_2 - w_1 = \pm\sqrt{(D - 27FG^2)/(9DG^2)}$$

For $\eta \neq 0$, equations (5.19), (5.20) can be reduced to the first or second Painlevé equation.

Physical meaning of the solutions obtained is of primary interest. There is no kink-type travelling waves in the problem stated above when the limiting values of strain are equal ($w_1 = w_2$) or if there is no energy exchange ($B = 0$). In these cases the travelling waves exist in the form of strain solitons or of cnoidal waves or in general, of the Weierstrass elliptic function. The algebraic kink-type solution for a rod with square nonlinearity and embedded into an active medium in terms of initial function $w(z)$ has the form:

$$w(z) = w_1 + (w_2 - w_1)exp(2\alpha z)(K + exp(\alpha z))^{-2}, \alpha < 0.$$

We can see that depending upon the type of the wave going from $(-\infty)$, transonic or subsonic, compressive or tensile, and upon the type of surrounding medium, the longitudinal physical strain component A is proportional to $\beta w(z)$ so that the combination of (sgnβ sgn w) defines the type of deformation of the rod. When $w_2 = -d^2/[25(a-b)]$ for $z \to -\infty$, the total unloading up to a vanishing strain is possible due to the existence of a kink wave.

When the rod is free of preliminary strain, the smooth kink wave leads to the generation of residual strain either behind the wave of compression (if $\beta > 0$) or behind the wave of tension (if $\beta < 0$). In these cases one should note that the values of amplitude and velocity of the impulse remain constant because the energy exchange (influx or dissipation) between the rod and the medium leads to the complete energy transition from the medium into the residual energy of nonlinearity elastic rod, namely, into the tension energy for $\beta < 0$ or into the compression energy for $\beta > 0$.

Thus an algebraic kink propagating along the wave guide embedded into an active medium plays a role of a trigger changing the initial state of an elastic wave guide into another one, depending upon the given properties of the wave guide and the medium and wave parameters.

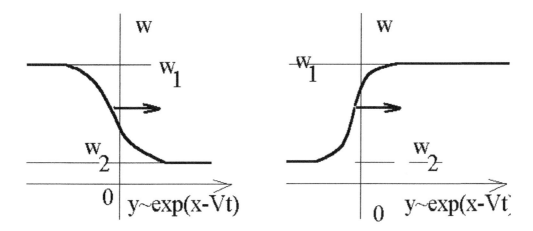

Figure 5.3: Kink deformation wave moving to the right may either: *a*) load the rod with a strain w_2, or: *b*) unload initially prestressed rod, depending on sgn β.

5.2.2 Waves in a rod in sliding contact with a surrounding medium

In this case the statement of the problem and the solutions will be changed dramatically. The presence of sliding resumes the influence of viscosity and plays a very specific role for solitary wave propagation in a rod embedded into an elastic medium. Most solutions will be periodic waves and solitons, not kinks. This problem was solved recently in (Porubov et al. 1998) and will be considered here.

Stresses on the lateral surface of an elastic wave guide, e.g., an elastic rod, may appear due to its interaction with the surrounding external medium, as in some technological devices. Various types of contact models can be used at the interface between the rod and the external medium. The full (strong) contact model is used when there is continuity of both normal and shear stresses and displacements. Alternatively, in a weak contact, friction may appear at the interface, hence a discontinuity in the shear stresses. Slippage provides with another form of contact at the interface, in which only the continuity of the normal stresses and displacements is assumed. Surface stresses may also arise due to the imperfect manufacturing of the lateral surface of the waveguide and are formally like the "surface tension" on the free surface of a liquid (see Biryukov et al. 1991; Nikolova, 1977).

We shall deal now with an appropriate description of long nonlinear strain waves propagating in an *elastic* cylindrical rod interacting with an external and different *elastic* medium by means of a *sliding* contact.

5.2.3   Statement and the solution to the problem

Consider an isotropic, axially infinitely extended elastic rod surrounded by another albeit different elastic medium, in which it may slide (like a nail in a piece of wood). We shall consider the propagation of longitudinal strain waves of small but finite amplitude in the rod when torsions are neglected. Assume, as before, the rod is made of a compressive Murnaghan material and the Hamilton principle expression should include the external forces work. It leads to the description of the rod similar to that used in Chapters 2 and 4.

Difference with the previous statement occurs if we consider the *linearly* elastic external medium surrounding the rod and having the density ρ_1 and elasticity governed by the Lamé coefficients (λ_1, μ_1).

Any perturbations caused by the wave propagation inside the rod are transmitted to the external medium through displacements and stresses, normal to the rod surface only, when contact with *slippage* is considered.

We shall also assume that disturbances vanish in the external medium far from the rod. The normal strains as well as the displacements inside the rod are smaller than those along the rod axis. Thus we assume that elastic displacements and strains are infinitesimal in the linear external medium. The displacement vector may be given as $\mathbf{U}_1 = (U_1, V_1, 0)$. Writing standard equations of linear elasticity for displacements in the external medium we have:

$$\rho_1 U_{1,tt} - (\lambda_1 + 2\mu_1)\, U_{1,zz} - (\lambda_1 + \mu_1)\left(V_{1,rz} + \frac{V_{1,x}}{r} \right) -$$

$$\lambda_1 \left(U_{1,rr} + \frac{U_{1,r}}{r} + V_{1,rx} + \frac{V_{1,x}}{r} \right) = 0 \qquad (5.23)$$

$$\rho_1 V_{1,tt} - (\lambda_1 + 2\mu_1)\left(V_{1,rr} + \frac{V_{1,r}}{r} - \frac{V_1}{r^2} \right) - \mu_1\, V_{1,xx} - (\lambda_1 + \mu_1) U_{1,rx} = 0 \qquad (5.24)$$

The following boundary conditions (b.c.) are imposed:

$$V \to 0, \qquad\qquad \text{at } r \to 0, \qquad\qquad (5.25)$$

$$V = V_1, \qquad\qquad \text{at } r = R, \qquad\qquad (5.26)$$

$$P_{rr} = \sigma_{rr}, \qquad\qquad \text{at } r = R, \qquad\qquad (5.27)$$

$$P_{rx} = 0,\ \sigma_{rx} = 0, \qquad\qquad \text{at } r = R, \qquad\qquad (5.28)$$

$$U_1 \to 0,\ V_1 \to 0 \qquad\qquad \text{at } r \to \infty. \qquad\qquad (5.29)$$

where P_{rr}, P_{rx} denote the components of the Piola - Kirchhoff stress tensor $\mathbf{P}$ (Lurie, 1980),

$$P_{rr} = (\lambda + 2\mu)\, V_r + \lambda\frac{V}{r} + \lambda\, U_x + \frac{\lambda + 2\mu + m}{2} U_r^2 + \frac{3\lambda + 6\mu + 2l + 4m}{2} V_r^2 +$$

$$(\lambda + 2l)\, V_r \frac{V}{r} + \frac{\lambda + 2l}{2}\frac{V^2}{r^2} + (\lambda + 2l)\, U_x V_r + (2l - 2m + n)\, U_x\frac{V}{r} +$$

$$\frac{\lambda + 2l}{2} U_x^2 + \frac{\lambda + 2\mu + m}{2} V_x^2 + (\mu + m)\, U_r V_x, \qquad\qquad (5.30)$$

$$P_{rx} = \mu \left(U_r + V_x\right) + \left(\lambda + 2\mu + m\right) U_r V_r + \left(2\lambda + 2m - n\right) U_r \frac{V}{r} +$$
$$\left(\lambda + 2\mu + m\right) U_x U_r + \frac{2m - n}{2} V_x \frac{V}{r} + \left(\mu + m\right) V_x V_r +$$
$$\left(\mu + m\right) U_x V_x \, . \tag{5.31}$$

The quantities σ_{rr} and σ_{rz} are the corresponding components of the linear stress tensor in the surrounding, external medium:

$$\sigma_{rr} = \left(\lambda_1 + 2\mu_1\right) V_{1,r} + \lambda_1 \frac{V_1}{r} + \lambda_1 U_{1,x} \tag{5.32}$$

$$\sigma_{rx} = \mu_1 \left(u_{1,r} + w_{1,x}\right) \tag{5.33}$$

The conditions (5.26)-(5.28) define the so-called *sliding* contact, while the longitudinal displacements U and U_1 remain free at the interface $r = R$.

The Piola- Kirchhoff tensor coincides with the linear stress tensor for infinitesimally small strains and was chosen among other finite strain tensors because it is defined in the reference configuration. The linear equations (5.23) and (5.24) are to be solved together with the boundary conditions (5.26), (5.28), (5.29), assuming that the displacement V at the interface is a given function of x and t, hence $V(x, t, R) = V(x, t)$. Then the linear *shear* stress σ_{rr} at the interface $r = R$ is obtained as a function of V and its derivatives, thus providing the dependence only on the rod characteristics in the right hand side of the b.c. (5.27). The same is valid for the elementary work done by external forces at $r = R$:

$$\delta A = 2\pi \int_{-\infty}^{\infty} \sigma_{rr} \, \delta V \, dx \, . \tag{5.34}$$

Satisfaction of the b.c. on the rod lateral surface yields the relationships between displacements and strains inside the rod, allowing to separate variables in the Lagrangian density and to derive the one nonlinear equation for long longitudinal waves using Hamilton's variational principle.

5.2.4 External stresses on the rod lateral surface

The linear problem (5.23), (5.24) will be solved with the boundary conditions (5.26), (5.28), (5.29). Aiming to study the *travelling* waves propagating along the axis of the rod, we assume that all variables depend only upon the phase variable $\theta = x - ct$, where c is now the phase velocity of the wave. Assuming that the unknown functions U_1, V_1 are

$$U_1 = \Phi_\theta + \Psi_r + \frac{\Psi}{r}, \qquad V_1 = \Phi_r - \Psi_\theta, \tag{5.35}$$

then Φ and Ψ satisfy the equations:

$$\Phi_{rr} + \frac{1}{r}\Phi_r + \left(1 - \frac{c^2}{c_l^2}\right)\Phi_{\theta\theta} = 0, \tag{5.36}$$

$$\Psi_{rr} + \frac{1}{r}\Psi_r - \frac{1}{r^2}\Psi + \left(1 - \frac{c^2}{c_\tau^2}\right)\Psi_{\theta\theta} = 0, \tag{5.37}$$

where c_l and c_τ are the velocities of the bulk longitudinal and shear linear waves in the external medium, respectively. They depend on the density and Lamé coefficients, $c_l^2 = (\lambda_1 + 2\mu_1)/\rho_1$, and $c_\tau^2 = \mu_1/\rho_1$.

To solve equations (5.36), (5.37)we introduce the Fourier transforms of Φ and Ψ:

$$\tilde{\Phi} = \int_{-\infty}^{\infty} \Phi \exp(-k\,\theta)\,d\theta, \quad \tilde{\Psi} = \int_{-\infty}^{\infty} \Psi \exp(-k\,\theta)\,d\theta$$

that reduces both equations (5.36), (5.37) to the Bessel equations :

$$\tilde{\Phi}_{rr} + \frac{1}{r}\tilde{\Phi}_r - k^2\alpha\,\tilde{\Phi} = 0, \tag{5.38}$$

$$\tilde{\Psi}_{rr} + \frac{1}{r}\tilde{\Psi}_r - \frac{1}{r^2}\tilde{\Psi} - k^2\beta\,\tilde{\Psi} = 0, \tag{5.39}$$

with $\alpha = 1 - c^2/c_l^2$, and $\beta = 1 - c^2/c_\tau^2$. The ratios between c, c_l and c_τ define the signs of α and β, hence three possible sets of solutions to the equations (5.38), (5.39) appear, vanishing at infinity, according to (5.29). Using the boundary conditions (5.26), (5.28),we obtain the following relationships for the Fourier images of normal stresses at the lateral surface $r = R$:

I) when $0 < c < c_\tau$:

$$\tilde{\sigma}_{rr} = \frac{\mu_1\,\tilde{V}}{1 - \beta}\left(\frac{2(\beta - 1)}{R} + \frac{k(1 + \beta)^2\,K_0(\sqrt{\alpha}kR)}{\sqrt{\alpha}\,K_1(\sqrt{\alpha}kR)} - \frac{4k\sqrt{\beta}K_0(\sqrt{\beta}kR)}{K_1(\sqrt{\beta}kR)}\right) \tag{5.40}$$

II) when $c_\tau < c < c_l$

$$\tilde{\sigma}_{rr} = \frac{\mu_1\,\tilde{V}}{1 - \beta}\left(\frac{2(\beta - 1)}{R} + \frac{k(1 + \beta)^2\,K_0(\sqrt{\alpha}kR)}{\sqrt{\alpha}\,K_1(\sqrt{\alpha}kR)} - \frac{4k\sqrt{\beta}J_0(\sqrt{-\beta}kR)}{J_1(\sqrt{-\beta}kR)}\right) \tag{5.41}$$

III) when $c > c_l$

$$\tilde{\sigma}_{rr} = \frac{\mu_1\,\tilde{V}}{1 - \beta}\left(\frac{2(\beta - 1)}{R} + \frac{k(1 + \beta)^2\,J_0(\sqrt{-\alpha}kR)}{\sqrt{-\alpha}\,J_1(\sqrt{-\alpha}kR)} - \frac{4k\sqrt{\beta}J_0(\sqrt{-\beta}kR)}{J_1(\sqrt{-\beta}kR)}\right) \tag{5.42}$$

where J_i and K_i ($i = 0, 1$) denote the corresponding Bessel functions.

We shall show, in the next subsection, that in the long wave limit the normal stress σ_{rr} has one and the same functional form at the lateral surface of the rod in all three cases (5.40) - (5.42).

The main difference in the stress (and strain) fields in the external medium is how they vanish at infinity, which depends on the monotonicity of decay of K_i and the oscillatory decay of J_i when $R \to \infty$. Note that the dependence of the strain wave behaviour on the velocities of bulk linear waves c_l, c_τ is known, in particular, for acoustic transverse Love waves propagating in an elastic layer superimposed on an elastic half-space, see (Jeffrey, Engelbrecht, eds. (1994), Parker, Maugin, eds. (1987), Mayer (1995)).

5.2.5 Derivation of strain-displacement relationships inside the rod

To solve the nonlinear problem inside the elastic rod, we have to simplify the relationships between longitudinal and shear displacements U and V. These relationships are obtained using conditions on the free lateral surface $r = R$, namely, the simultaneous absence of the tangential stresses and the continuity of the normal ones. The approach is similar to that used before, however the coefficients will be different.

Dealing with *elastic* strain waves with sufficiently small magnitude $B \ll 1$, and a *long* wave length Λ relative to the rod radius R, $R/\Lambda \ll 1$, we suggest the balance between (weak) nonlinearity and (weak) dispersion as for a rod with a free lateral surface. Then the small parameter of the problem

$$\varepsilon = B = (R/\Lambda)^2 \ll 1. \tag{5.43}$$

The linear part of longitudinal strain along the rod axis, C_{xx}, is U_x. Then choosing Λ as a scale along x, one gets $B\Lambda$ as a scale for displacement U. Similarly, the linear part of transverse strain, C_{rr}, is V_r, and the natural scale for displacement V is BR, where R is a length scale along the rod radius. Then with $|kR| \ll 1$ in (5.40) - (5.42), we have a power series expansion in kR. It allows us to obtain analytically an inverse Fourier transform for σ_{rr} and to write the conditions (5.27), (5.28) at $r = 1$ in dimensionless form as:

$$(\lambda + 2\mu)\, V_r + (\lambda - k_1)\frac{V}{r} + \lambda\, U_x + \frac{\lambda + 2\mu + m}{2} U_r^2 +$$
$$\varepsilon(\frac{3\lambda + 6\mu + 2l + 4m}{2} V_r^2 + (\lambda + 2l)\frac{V}{r} V_r + \frac{\lambda + 2l}{2}\frac{V^2}{r^2} + (\lambda + 2l)\, U_x V_r +$$
$$(2l - 2m + n)\, U_x \frac{V}{r} + (\mu + m)\, U_r V_x + \frac{\lambda + 2l}{2} U_x^2 - k_2\, r\, V_{xx}) +$$
$$\varepsilon^2 \frac{\lambda + 2\mu + m}{2} V_x^2 = O(\varepsilon^3), \tag{5.44}$$

$$\mu U_r +$$
$$\varepsilon \left(\mu\, V_x + (\lambda + 2\mu + m)\, U_r V_r + (2\lambda + 2m - n)\, U_r \frac{V}{r} + (\lambda + 2\mu + m)\, U_x U_r \right) +$$
$$\varepsilon^2 \left(+\frac{2m - n}{2}\frac{V}{r} V_x + (\mu + m)\, V_x V_r + (\mu + m)\, U_x V_x \right) = O(\varepsilon^3) \tag{5.45}$$

Obviously, $V \equiv V(x,t)$, $V_{xx} \equiv V_{xx}(x,t)$ at the rod lateral surface. Moreover, for $0 < c < c_\tau$:

$$k_1 = -2\mu_1, \quad k_2 = \frac{\mu_1 c^2 (\gamma - \log 2)}{c_\tau^2}, \tag{5.46}$$

while for $c_\tau < c < c_l$:

$$k_1 = \frac{2\mu_1 (4c_\tau^2 - c^2)}{c^2}, \quad k_2 = \frac{\mu_1 c_\tau^2}{c^2}\left(1 - \frac{c^2}{c_\tau^2} + (2 - \frac{c^2}{c_\tau^2})^2(\gamma - \log 2)\right), \tag{5.47}$$

and for $c > c_l$:

$$k_1 = \frac{2\mu_1[c^2(c_\tau^2 - c_l^2) + 3c_l^2 c_\tau^2 - 4c_\tau^4]}{c_\tau^2(c_l^2 - c^2)}, \quad k_2 = \frac{\mu_1 c^2}{4c_\tau^2}. \tag{5.48}$$

where $\gamma = 0.5772157$ is Euler's constant.

The unknown functions U, V will be found in power series of ε:

$$U = U_0 + \varepsilon U_1 + \varepsilon^2 U_2 + \dots, \quad V = V_0 + \varepsilon V_1 + \varepsilon^2 V_2 + \dots. \tag{5.49}$$

Substituting (5.49) in (5.44), (5.45), and equating to zero all terms of the same order of ε, we find again that the plane cross-section hypothesis and Love's relationship both are valid in the leading order only:

$$U_0 = U(x,t), \quad V_0 = rC U_x, \tag{5.50}$$

while the coefficient is different

$$C = \frac{\lambda}{k_1 - 2(\lambda + \mu)}. \tag{5.51}$$

In order $O(\varepsilon)$ we obtain:

$$U_1 = -r^2 \frac{C}{2} U_{xx}, \quad V_1 = r^3 D U_{xxx} + r Q U_x^2, \tag{5.52}$$

with coefficients:

$$D = \frac{\lambda(\lambda + 2k_2)}{2(k_1 - 2(\lambda + \mu))(2(2\lambda + 3\mu) - k_1)} \tag{5.53}$$

$$Q = \frac{1}{k_1 - 2(\lambda + \mu)} \left(\frac{\lambda + 2l}{2} + C(\lambda + 4l - 2m + n) + C^2(3\lambda + 3\mu + 4l + 2m) \right) \tag{5.54}$$

The higher order terms in the series (5.49) may be obtained in a similar way, but are omitted here being unnecessary to obtain an evolution equation for the strain waves.

5.2.6   A nonlinear evolution equation for longitudinal strain waves along the rod and its solution

To derive the equation for the strain waves along the rod we substitute (5.49) into the potential deformation energy Π and obtain in dimensionless form

$$\Pi = a_1 U_x^2 + \varepsilon^2 \left[a_2 r^2 U_x U_{xxx} + a_3 U_x^3 \right] + O(\varepsilon^4), \tag{5.55}$$

where

$$a_1 = \frac{\lambda + 2\mu}{2} + 2\lambda C + 2(\lambda + \mu)C^2,$$

$$a_2 = -\frac{\lambda + 2\mu}{2} C - \lambda C^2 + 4\lambda D + 8(\lambda + \mu) C D,$$

$$a_3 = \frac{\lambda + 2\mu}{2} + \lambda C + \lambda C^2 + 2(\lambda + \mu)C^3 + 2Q\left[\lambda + 2(\lambda + \mu)C\right] +$$

$$l\left[\frac{1}{3} + 2C + 4C^2 + \frac{8}{3}C^3\right] + m\left[\frac{2}{3} - 2C^2 + \frac{4}{3}C^3\right] + nC^2.$$

For the kinetic energy we have:

$$K = \frac{\rho_0}{2}\left[U_t^2 - \varepsilon^2 r^2 C(U_t U_{xxt} - CU_{xt}^2)\right] + O(\varepsilon^4) \tag{5.56}$$

Substituting (5.55), (5.56) and (5.34) into the Hamilton's variational principle, we obtain the following DDE for a longitudinal strain wave $u = U_x$:

$$u_{tt} - b_1\, u_{xx} - \varepsilon^2\left(b_2\, u_{xxtt} + b_3\, u_{xxxx} + b_4\, (u^2)_{xx}\right) = 0, \tag{5.57}$$

similar to the equation studied in Chapters 2-4 derived for a free rod , however, with different coefficients:

$$b_1 = \frac{2(a_1 - k_1 C^2)}{\rho_0}, \quad b_2 = \frac{C(1+C)}{2},$$

$$b_3 = \frac{a_2 - 2C(k_2 C + 2 k_1 D)}{\rho_0}, \quad b_4 = \frac{3(a_3 - k_1 C Q)}{\rho_0}. \tag{5.58}$$

It means that the *sliding contact* problem is, in a sense, conservative and admits, in particular, a travelling classical solitary wave as an exact solution.

Note that the coefficients depend now upon the wave velocity, c, due to (5.46)-(5.48). Terms of order $O(\varepsilon^4)$ have been neglected, when deriving equation (5.57), therefore we assume $c^2 = c_0^2 + \varepsilon^2 c_1 + \dots$ and consider the coefficients $b_2 - b_4$ depending on c_0 only, while the coefficient b_1 may depend also on c_1 as $b_1 = b_{10}(c_0) + \varepsilon^2 b_{11}(c_0, c_1)$. Then the solitary wave solution possesses the form:

$$v = A\, m^2 \cosh^{-2}(m\,\theta), \tag{5.59}$$

with

$$A = \frac{6(b_{10}b_2 + b_3)}{b_4}. \tag{5.60}$$

To a leading order the phase velocity is obtained from the equation

$$c_0^2 = b_{10}(c_0), \tag{5.61}$$

and for the function c_1 we get the equation

$$c_1 = b_{11} + 4k^2(b_{10}b_2 + b_3), \tag{5.62}$$

where the wave number k remains a free parameter.

5.2.7 The influence of the external medium on the propagation of the strain soliton along the rod

The main objective is to estimate the influence of the external medium on the solitary wave propagation along the rod. First of all, we have to solve equation (5.61) for all possible cases (5.46) -(5.48). As ε must not exceed the yield point of the elastic material (its usual value is less than 10^{-3}) we have to compare with c_l and c_τ the values obtained for c_0, rather than for c.

If the phase velocity is small, as in the case (5.46), the velocity c_0 is obtained from (5.61) as

$$c_0^2 = \frac{(3\lambda + 2\mu)\mu + \mu_1(\lambda + 2\mu)}{\rho_0(\lambda + \mu + \mu_1)}. \tag{5.63}$$

It is always *higher* than the wave phase velocity in a free rod, hence the contact with an external medium in this case leads to the soliton acceleration.

For the model governed by (5.47), the equation (5.61) yields an algebraic relation:

$$c_0^4 - \frac{(3\lambda + 2\mu)\mu + \mu_1(\lambda + 2\mu) + 4\mu_1\rho_0 c_\tau^2}{\rho_0(\lambda + \mu + \mu_1)} c_0^2 + \frac{4\mu_1 c_\tau^2(\lambda + 2\mu)}{\rho_0(\lambda + \mu + \mu_1)} = 0 \tag{5.64}$$

Finally, for the model (5.48), equation (5.61) provides:

$$c_0^4 - \frac{(3\lambda + 2\mu)\mu c_\tau^2 + (c_\tau^2 - c_l^2)\mu_1(\lambda + 2\mu) + 4\mu_1\rho_0 c_\tau^4 + c_\tau^2 c_l^2 \rho_0(\lambda + \mu - 3\mu_1)}{\rho_0(c_l^2\mu_1 - c_\tau^2(\lambda + \mu + \mu_1))} c_0^2 +$$

$$\frac{c_\tau^2 c_l^2[3\mu_1(\lambda + 2\mu) - \mu(3\lambda + 2\mu)] - 4\mu_1 c_\tau^4(\lambda + 2\mu)}{\rho_0(c_l^2\mu_1 - c_\tau^2(\lambda + \mu + \mu_1))} = 0 \tag{5.65}$$

Table 5.1 contains some quantitative estimates for a polystyrene rod and Table 5.2 for a lead rod, respectively, both embedded in different external media.

Table 5.1. Phase velocities of waves in a polystyrene rod embedded in different media.

velocity$\times 10^{-3}$, $m/\,sec$	c_τ	c_l	c_{01}	c_{02}	c_{03}	
material						**model**
quartz	3.78	6.02	2.06	2.1 or 7.15	2.13 or 5.77	I
iron	3.23	5.85	2.08	2.1 or 6.32	2.11 or 5.15	I
copper	2.26	4.7	2.07	2.11 or 4.33	2.12 or 3.68	I, II
brass	2.12	4.43	2.06	2.11 or 4.02	2.12 or 3.45	I, II
aluminium	3.08	6.26	2.05	2.11 or 5.75	2.13 or 4.97	I, II
lead	1.09	2.41	2.01	−	1.83 or 2.06	-

The quantities c_{01}, c_{02} and c_{03} denote velocities calculated according to the relationships (5.63), (5.64) and (5.65), respectively. Comparing velocities c_{0i} relative to c_τ and c_l we can justify the applicability of cases (5.46)- (5.48). This is notified by symbols I-III, respectively, in the last column of Tables 5.1 and 5.2.

Indeed, the model (5.46) is better for the contact with a polystyrene rod, while no solitary wave may propagate when the external medium is lead. However, a solitary wave may propagate along a lead rod embedded in a polystyrene external medium, as it follows from Table 5.2.

Table 5.2. Phase velocities of waves in a lead rod embedded in different external media.

velocity$\times 10^{-3}$, $m/$ sec	c_τ	c_{lDe}	c_{01}	c_{02}	c_{03}	
material						**model**
quartz	3.78	6.02	2.06	2.55 or 4.39	7.51	I,II,III
iron	3.23	5.85	2.2	2.47 or 4.91	2.73 or 4.81	I, II
copper	2.26	4.7	2.11	–	–	I
brass	2.12	4.43	2.08	–	–	I
aluminium	3.08	6.26	2.03	–	–	I
polystyrene	1.01	2.1	1.83	0.38 or 1.81	1.84 or 2.06	II, III

We should note a remarkable difference from a free rod problem: there are pairs of materials, for which two or even all three variants of sliding contact admit the solitary wave propagation. Thus the balance between nonlinearity and dispersion may be achieved at different phase velocities of the strain nonlinear waves. This result is of importance for experiments in generation of strain solitary waves in a rod embedded in an external elastic medium.

Therefore, strain solitary waves can propagate only with velocities from the intervals close to the value c_{0i}. Note that the solitary wave is a bulk (density) wave inside the rod, and simultaneously, it is a surface wave for the external medium. Then, an important difference appears relative to long nonlinear Rayleigh surface waves in cartesian coordinates: in our case more than one velocity interval exists where solitary waves may propagate. The main difference between modes lies in the different rate of wave decay in the external medium that follows from the different behavior of Bessel's functions at large values of their arguments.

Another question of considerable interest is in the influence of the elasticity of external surrounding medium on the existence of either compression or tensile longitudinal strain localized waves.

Using the data from Table 5.1 to compute the value of A (5.60) for a polystyrene rod, it yields that its sign may change according to the values of the parameters of the material used for the external medium. Therefore the (*strain !*) soliton amplitude (5.59) may change the sign. The amplitude is negative for a free lateral surface rod and it remains negative if the external medium is, e.g., quartz, brass, copper or iron. However, the sign changes if $c_0 = c_{02}$ as it will be for the external medium made of aluminium. Therefore, one can anticipate, in particular, that for a rod embedded in aluminium, an initial pulse with velocity close to c_{02} may transform only to *a tensile* soliton while an initial pulse with velocity close to c_{01} tends to a *compression* soliton.

Finally, let us consider the influence of sgn c_1 in (5.62). For case I, $b_{11} = 0$, hence the sign is defined by the sign of the quantity $(b_{10}b_2 + b_3)/b_4$. For polystyrene it is, generally, negative for all the external media in Table 1, while for a free lateral

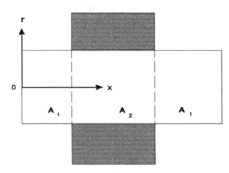

Figure 5.4: Rod partially embedded in an external elastic medium.

surface it is positive. Thus, the velocity c of a *nonlinear* wave in a rod embedded in an external medium is lower than the *linear* wave velocity, c_0, while for a free surface rod nonlinear waves propagate faster than linear waves. On the other hand, the nonlinear wave velocity c in a polystyrene rod embedded in external medium is higher than the linear wave velocity for a rod with free lateral surface, $c^* = \sqrt{E/\rho_0}$.

5.2.8 A numerical simulation of unsteady strain wave propagation

Recent numerical simulation of unsteady nonlinear wave processes in elastic rods with *free lateral surface* shows that for $A < 0$ only initial compression pulses provide a solitary wave (5.59) or a wave train (see Figure 3 in Samsonov, 1997, and Chapter 6 below), while tensile initial pulses do not become localized and are destroyed by dispersion. On the contrary, for $A > 0$ only tensile strain solitary waves may appear, and initial compression pulses are destroyed.

We consider the problem in which the rod lateral surface is partly free along the axis and the other part is subjected to a sliding contact with an external elastic medium, as it is shown in Figure 5.4. Then the nonlinear strain wave propagation is described in each part of the rod by its particular version of equation (5.57). Matching of solutions is provided by the continuity of strains and its derivatives. Assume that for the free surface part ($k_1 = 0$, $k_2 = 0$) $A = A_1$, $m = m_1$, while for the embedded one, $A = A_2$, $m = m_2$. Let the initial solitary wave (5.59) move from left to right far from the embedded part, which is supposed to be undeformed at the initial moment. It was found in Chapter 2 that the mass M conservation law in the form

$$\frac{d}{dt}M = 0, \ M = \int_{-\infty}^{\infty} u \, dx \tag{5.66}$$

is satisfied by equation (5.57). Then using (5.59) and (5.60) we obtain the estimation for the mass M_1

$$M_1 = 2 A_1 m_1, \tag{5.67}$$

The wave evolution along the embedded part depends on the ratio between A_1 and A_2. Similar to the unsteady processes inside a rod with the free lateral surface, an initial strain solitary wave will be destroyed in an embedded part, if $sgn A_2$ differs from $sgn A_1$. Otherwise another solitary wave or a wave train will appear. When the initial pulse is not massive enough it was found, that only one new solitary wave appears but there is an oscillatory decaying tail behind it. However, the contribution of the tail to the mass M is negligibly small relative to the solitary wave contribution, hence

$$M_2 = 2 A_2 m_2. \tag{5.68}$$

Comparing M_1 and M_2, according to (5.66) it follows

$$A_1 m_1 = A_2 m_2. \tag{5.69}$$

Therefore, if $A_2 < A_1$ the amplitude of the solitary wave increases while its width, proportional to m^{-1}, decreases, hence there is focusing of the solitary wave. On the contrary, when $A_2 > A_1$ attenuation of the solitary wave is provided by the simultaneous decrease of the amplitude and the increase of the wave width.

Numerical simulations confirm our theoretical estimates. In Figure 5.5 the evolution of a strain *tensile* solitary wave is shown in a rod, having a central part embedded in an external medium. The value of A_2 in the central part II is positive but smaller than the value of A_1 in the surrounding free lateral surface parts I and III: $A_1 > A_2 > 0$. In the embedded part II (Figure 5.5(b)) the solitary wave amplitude exceeds the amplitude of the initial solitary wave in Figure 5.5(a), while its width becomes smaller than that of the initial wave. Therefore an increase in amplitude of the elastic strain solitary wave is possible even in a *uniformly* elastic rod due to interface interaction. This may exceed the yield point inside the elastically deformed rod, and cracks or plasticity zones may appear.

In our case the deformations of the wave's front and back are equal. In the problem of the strain soliton focusing in a tapered rod (Chapter 4) both theory and experiments show the steepness of the wave front along with smoothening of its back. Moreover, a *plateau* transforms in the tail of the solitary wave. These differences result from the absence of mass (and energy) conservation for strain solitary waves in the tapered rod, moreover the decay of mass can be calculated.

In the problem under consideration the solitary wave does not lose mass, M, hence its original shape is recovered in the part III in Figure (5.5,c,d). One can see that an oscillatory tail of the solitary wave in Figure 5.5(d) is less than the tail in Figure 5.5(c), in good agreement with (5.69).

When $A_2 > A_1 > 0$, an initial tensile strain solitary wave, see Figure 5.6(a), is dramatically attenuated as soon as it enters the embedded area, see Figure 5.6(b), its amplitude decreases while its width becomes larger. Again both the recovery of the initial wave shape and the damping of its tail may be observed in the third part of a rod with free lateral surface, part III in Figure 5.6(c,d).

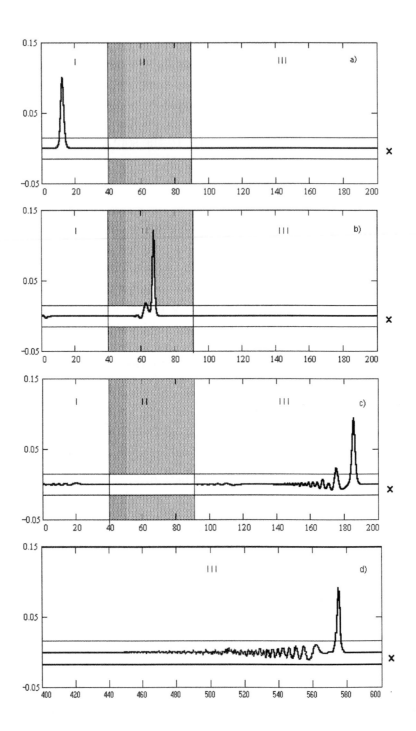

Figure 5.5: Focus and recovery of the tensile strain soliton.

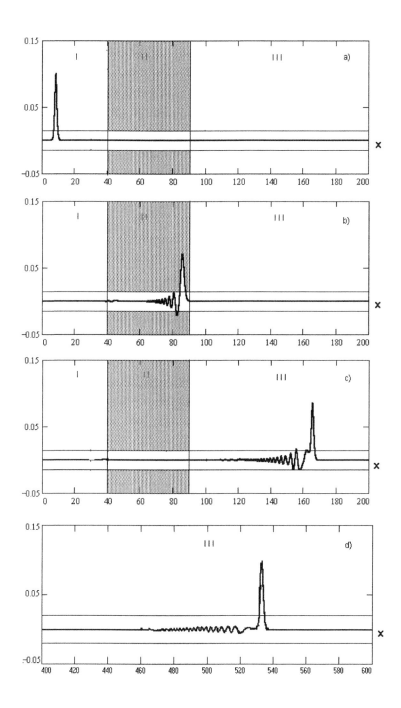

Figure 5.6: Decay and recovery of the soliton after the embedding area.

Consider now the case of *different* signs of the influence factors A_i and assume that $A_1 > 0$ on both free surface parts. One can see in Figure 5.7 how an initial tensile solitary wave shown in Figure 5.7(a), is destroyed in the embedded part II, Figure 5.7(b), in agreement with our previous results on the unsteady processes occurring for a free surface rod. However, a strain wave is localized again in the third part of a rod with free lateral surface, Figure 5.7(c), and eventually recovers its initial shape in Figure 5.7(d). The damping of the tail behind the solitary wave is observed also. Therefore we have shown that compression as well tensile initial pulses both may produce the localized strain solitary waves in a rod partly embedded in an external elastic medium with sliding.

Moreover, the amplitude of the solitary wave generated in such a manner may increase comparing with the magnitude of the initial pulse. This is shown in Figures 5.8, 5.9, where $A_1 < 0$, $A_2 > 0$ and $|A_1| < A_2$. One can see in Figure 5.8(a) how an initial localized rectangular tensile pulse is destroyed in the free surface part immediately upon generation as in Figure 5.8(b). However, a waves train of solitons appears when the partially destroyed strain wave moves to the embedded part, having "friendly" parameters as in Figures 5.8(c,d). Moreover, the amplitude of the first soliton in Figure 5.8(d) exceeds the magnitude of the initial rectangular pulse.

In the absence of the surrounding external medium this rod does not support tensile solitary wave propagation at all, and a strain wave is delocalized as shown in Figure 5.9.

5.2.9 Applications of the theory

Surface tension-like effects

The solitary strain wave theory developed may be applied to the study of surface tension-like effects in solids, e.g., if there are imperfections on the rod surface, see, e.g., (Biryukov et al. 1991). It was experimentally found (Nikolova, 1977) that the stresses caused by the surface effects, may be rather large; the theory (Biryukov et al. 1991) shows that surface stresses, acting on the lateral surface of an elastic body, may be modelled using normal stresses in the form

$$\sigma_{rr} = \alpha_{eff}\, v_{xx}, \tag{5.70}$$

with α_{eff} being a surface tension-like coefficient. In this case the boundary conditions (5.44), (5.45) are valid with $k_1 = 0$, $k_2 = \alpha_{eff}$. Thus our theory may be *formally* extended to take into account the influence of surface tension-like effects on the propagation of strain solitary waves. This "surface tension" does not alter the phase velocity c though it may effect the sign of the wave amplitude. Although the problem of obtaining meaningful values of the "surface tension" coefficient in solids is far from being solved, the data given by Nikolova (1977) for some materials seems to be reliable. The theory developed here may be used for the determination of the surface tension-like coefficient. Indeed, the expression $(b_{10}b_2 + b_3)/b_4$ contains α_{eff}, hence, by measuring solitary wave parameters in a rod with different surface roughness one can obtain the corresponding values of α_{eff}. Accordingly, an estimation of

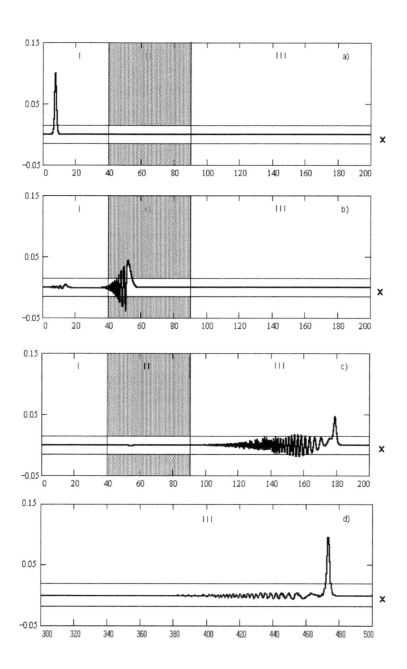

Figure 5.7: Delocalization and recovery of the strain soliton due to embedding.

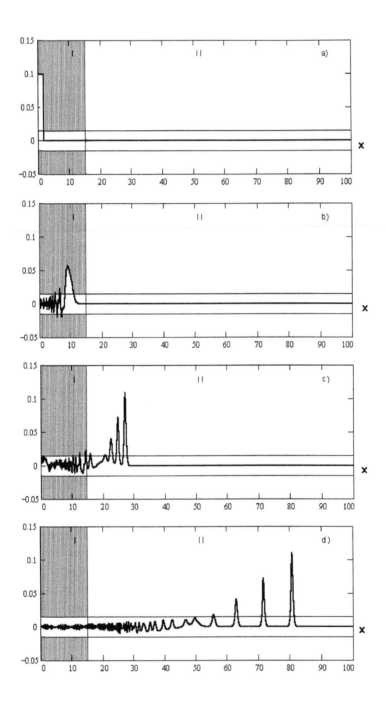

Figure 5.8: Generation of the tensile strain solitary wave train in a rod. A tensile wave cannot propagate in this rod in the absence of embedding, as in part I.

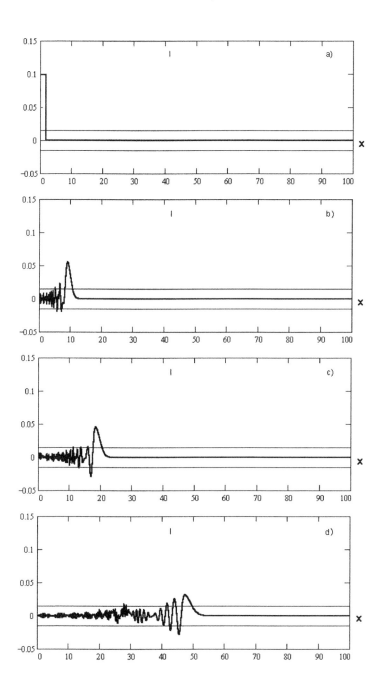

Figure 5.9: Decay and dispersion of the tensile pulse in the absence of the external medium.

the influence of the surface tension on the solitary wave parameters is useful to apply the theory to nondestructive testing, because a bulk strain solitary wave (5.59) keeps its shape independently of the lateral surface roughness, while the wave parameters (amplitude, velocity etc.) contain information about it.

Murnaghan's moduli

The isotropic third order Murnaghan's moduli (l, m, n) are not known for many materials or the data are not reliable. The third order crystalline moduli have been measured for some materials (see Bogardus, 1965; Lurie, 1980; Frantsevich et al. 1982). In a paper by Thurston and Brugger (1964) it was proposed to use them to obtain the isotropic bulk moduli. For cubic crystals the relationships are:

$$c_{112} = 2l, \; c_{166} = m, \; c_{456} = n/4, \; c_{123} = n - 2m + 2l,$$

$$c_{144} = m - n/2, \; c_{111} = 4m + 2l. \tag{5.71}$$

where c_{ijk} denotes the corresponding third order crystalline elastic modulus for cubic crystals. However, independent measurements of isotropic moduli for some materials do not satisfy these analytical relationships. For instance, for aluminium and molybdenum, for which both the Murnaghan moduli and the crystalline cubic moduli are known after (Frantsevich et al. 1982) we can estimate the discrepancy. Using equations (5.71) we calculated the deviations ,

$$\delta_1 = \left| \frac{c_{112} - 2l}{c_{112}} \right|; \; \delta_2 = \left| \frac{c_{166} - m}{c_{166}} \right|; \; \delta_3 = \left| \frac{c_{456} - 0.25n}{c_{456}} \right|;$$

$$\delta_4 = \left| \frac{c_{123} - n + 2m - 2l}{c_{123}} \right|; \; \delta_5 = \left| \frac{c_{144} - m + 0.5n}{c_{144}} \right|; \delta_6 = \left| \frac{c_{111} - 4m - 2l}{c_{111}} \right|.$$

presented in Table 5.3:

Table 5.3. Deviations in *percents* from (5.71)

material	δ_1	δ_2	δ_3	δ_4	δ_5	δ_6
aluminium	60	15	111	92	208	50
molibdenum	27	52	115	305	604	39

There is a continuous interest in the direct measurement of the Murnaghan moduli for both physics and technology purposes. Our theory gives one of the possible ways. It follows from (5.58) that

$$b_4 = q_0 + q_1 \, l + q_2 \, m + q_3 \, n, \tag{5.72}$$

is a linear combination of Murnaghan's moduli, with

$$q_0 = \frac{H}{2\rho_0[2(\lambda + \mu) - k_1^2]^4},$$

where

$$H = 48\mu(\lambda + \mu)^3(3\lambda + 2\mu) + 3k_1^4(\lambda + 2\mu) - 3k_1^3(\lambda + 4\mu)(5\lambda + 4\mu)$$
$$+ 12k_1^2(3\lambda + 2\mu)(\lambda^2 + 6\lambda\mu + 6\mu^2) - 6k_1(\lambda + \mu)(3\lambda + 2\mu)(\lambda^2 + 16\lambda\mu + 16\mu^2);$$

and

$$q_1 = \frac{(2\mu - k_1)(k_1^2 + k_1(\lambda - 4\mu) + 4\mu(\lambda + \mu))}{2\rho_0[2(\lambda + \mu) - k_1^2]^4};$$

$$q_2 = \frac{G}{\rho_0[2(\lambda + \mu) - k_1^2]^4};$$

where

$$G = 2(k_1 - 3\lambda - 2\mu)(k_1^3 - k_1^2(5\lambda + 6\mu) + k_1(3\lambda^2 + 20\lambda\mu + 12\mu^2) - 4\mu(3\lambda^2 + 5\lambda\mu + 2\mu^2));$$

$$q_3 = \frac{6\lambda^2(\lambda + \mu - k_1)}{\rho_0[2(\lambda + \mu) - k_1^2]^3}.$$

The coefficients $q_0 - q_4$, b_4 are functions of the well known Lamé coefficients and densities of the rod and the external medium. The coefficient b_4, in addition, depends upon the solitary wave amplitude (5.59), (5.60). Hence, the equation (5.72) may be considered as a linear inhomogeneous algebraic equation for the Murnaghan moduli (l, m, n). Making measures and calculations in the solitary wave propagation in a rod embedded in three different external media, we may have three equations and obtain the values of l, m, n. The necessary and sufficient condition for a nontrivial solution is the non-zero value of the determinant of the system. Calculations for several elastic materials show that it usually does not vanish. However for the time being, the problem of sufficiently accurate measurement of the solitary wave amplitude in experiments is hardly solved with given accuracy, and in particular, the width or wave length of the solitary wave does not have an appropriate definition, so it cannot be clear defined in experiments. Therefore, it is better to find the periodic waves trains. The DDE (5.57), indeed, admits such a solution in the form of a cnoidal wave

$$u = \frac{6(b_{10}b_2 + b_3)}{b_4} k^2 \left[1 - \frac{E}{K} - \kappa^2 + \kappa^2 cn^2 \left(k\,\theta \,|\kappa \right) \right] \tag{5.73}$$

where K, E and κ are the complete elliptic integrals of the first and second kinds and Jacobi functions modulus, respectively. Using the cnoidal wave solution, it is not a problem to define all wave characteristics with given accuracy, and it follows from (5.73) the possibility of determination of Murnaghan's moduli.

Unfortunately, no experimental data are available up-to-date, concerning the generation of such a wave even in a rod with a free lateral surface, and it will provide a challenge for experimentalists.

5.2.10 Conclusions

We proposed a theory to study the nonlinear longitudinal strain waves in an elastic rod embedded in another external elastic medium with sliding contact. First, relationships were obtained for the normal stresses acting on the rod lateral surface. Then, in the long wave limit we derived the nonlinear evolution equation for strain waves in a rod, and an exact solitary wave solution was obtained. The analysis of the solution allowed us to conclude that the influence of the external medium defines an interval of phase velocities in which a solitary wave can propagate.

In contrast to *surface* wave propagation description in cartesian coordinates for wave guides, where only one wave velocity is possible, here we have two or even three intervals of *"allowed"* velocities. Moreover, depending on the elasticity of the surrounding external medium the longitudinal bulk strain wave in a rod may provide either a tensile or a compression wave.

An appearance and evolution of the nonlinear wave in a rod partly embedded in an elastic external medium was studied in numerical simulation. Focusing, attenuation or delocalization of a strain solitary wave are observed in such a case, depending on the elasticity of the external medium. Moreover, in each of these cases we found a recurrence of the original solitary wave when moving out of the embedded area. As a result of wave focusing, the strain exceeding the yield point of the elastic rod material may occur, as well as possible localization of either compression or tensile pulses. All these properties could be useful to design the elastic structures or to establish criteria to estimate properly their durability and fracture threshold.

A generalization of the theory has been proposed to formally account for surface tension-like effects on the evolution of long nonlinear strain waves. This generalisation may also be of interest for usage of nonlinear waves as new instrument in the nondestructive testing. Finally, we have shown how the theory may provide a direct determination of the Murnaghan third order isotropic elastic moduli of the material by measuring the parameters of the solitary wave propagating along the rod.

5.3   Nonlinear waves in the layer upon the elastic half space

Another nonlinear wave propagation problem, based on the appropriate modelling of contact of two different elastic media, occurs in the study of the long nonlinear strain waves propagation problem in the thin elastic layer superimposed on elastic half-space. In further consideration we follow mostly the paper by Porubov and Samsonov (1995), and several misprints will be corrected; the preliminary results were presented by Porubov and Samsonov (1994).

5.3.1   Physical background

Even recently the *envelope* waves propagation problems in layered half-space were considered as a rule, see, e.g., Maradudin (1987), Maugin and Hadouaj (1990), where only SH-wave (the Love wave) evolution without change of shape was studied. The nonlinear Schrödinger equation was derived by Kalyanasundaram, (1987), in order to describe a surface acoustic soliton propagation, and the Rayleigh and Love waves interactions were studied by means of solutions of coupled system of nonlinear Schrödinger equations without dispersion. The theory of long nonlinear wave propagation is not well developed for layered and nonlinearly elastic structures, and we have to refer to a few papers by Ewen et al. (1982), Cho and Miyagawa (1993) and Sakuma and Nishiguchi (1990).

The main interest in the problem is in the fact that despite of inconsistent theory such waves were observed and studied recently by Dyakonov et al. (1988a), Cho and Miyagawa (1993) and Nayanov (1986), e.g., in the experiments on the long finite amplitude surface acoustic wave propagating along a thin superconductive metal film on $LiNbO_3$ substrate , see Dyakonov et al. (1988b). It was found that elastic deformation, arising due to intensive acoustic wave propagation, strongly influenced the width of the superconductivity threshold, moreover, the shape of the observed wave just before a layered area was surprisingly similar to the Benjamin-Ono periodical wave solution. It is to be noted that this equation was proposed by Ewen et al. (1982) for the model of the surface acoustic wave propagation. However the derivation was based on phenomenology, not on the consideration of the basic nonlinear elasticity equations. That is why the analytical relationships were not found there for the equation coefficients as well as the influence of the layer-half-space contact type on the equation form was not clarified.

We consider here the problem of long strain waves propagation in the nonlinear elastic thin layer superimposed on nonlinear elastic half-space, both of which are assumed to be isotropic for simplicity.

Again, as in the rod problems two different interface contacts are to be studied.

According to the first of them, the contact provides by means of the continuity of normal stresses and displacements components only, while both the layer and the half-space shear stresses components are supposed to be zero at the interface, and the displacement tangential components are free. This statement of the problem is called as the contact with slippage and closely related to the problem of an elastic plate lying on an elastic basement.

The second model to be considered is based on the complete continuity conditions for all stresses and displacements components at the interface. We will derive the basic equations for description of the longitudinal nonlinear strain waves evolution in both cases.

The Benjamin-Ono (B-O) equation arises for long waves, indeed, in the first case, and it will be established that the 3d order elastic moduli only of a layer (but not of a half-space) include in the equation nonlinear term coefficient. Studying the envelope waves propagation, we derive the nonlinear Schrödinger equation (NLS). In the full contact model we will obtain the more complicated integro-differential equation, in which the nonlinear term coefficients now depend on the 3d order elastic moduli of the half-space only. Moreover, the last equation is unlikely to possess the solutions similar to the B-O ones, and again we could obtain the NLS equation to describe the envelope wave solution in the framework of full contact model.

Therefore it seems to be that the periodical and solitary waves governed by the Benjamin-Ono equation may propagate only when the layer slippage has been taken into account in the statement of the problem. This may be important for possible application to description of acoustic waves localization that was proposed in (Dyakonov et al. (1988a)) as a cause of the width variations of superconductive threshold area in metal films on $LiNbO_3$ substrate. The linear wave analysis will be provided also in order to get the improved (in comparison with those obtained by Tiersten (1969)) dispersion relation for long longitudinal strain waves propagation in layered half-space.

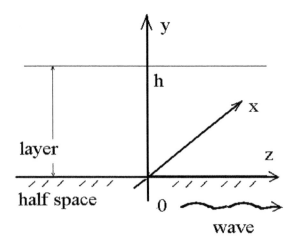

Figure 5.10: Schematic of the layered half space

The results obtained allows us to conclude that the Korteveg-de Vries equation is unlikely to govern such a wave propagation problem, contrary to the statement that was proposed by Ewen et al. (1982), Cho and Miyagawa (1993), Nayanov (1986).

5.3.2 Basic equations

We consider the wave propagation in the thin elastic layer superimposed on an elastic half-space. A plane (x, z) of the cartesian coordinate system (x, y, z) is assumed to coincide with the surface boundary between the layer and the half-space, and the layer occupies the interval $0 \leq y \leq h$. Let us assume that the waves propagate along the z-axis, and all components of displacements vector $\mathbf{U}(\mathbf{r}, t) = (0, V_i(y, z, t), W_i(y, z, t))$ are independent on x.

Here and in the following a subscript value $i = 1$ corresponds to the layer and $i = 2$ - to the half-space parameters, while a letter subscript other then i will denote the differentiation with respect to the corresponding variable. Linear elastic features of the layer and of the half-space are characterized by their densities ρ_i and the Lamé constants λ_i and μ_i while the nonlinear elastic features is suggested to be described in addition by the third order (the Murnaghan) moduli l_i, m_i, n_i.

We shall start with the basic statement of the elasticity problem as above, write the Lagrangian density per volume unit $\mathcal{L}_i$ in each elastic medium as the difference of the kinetic energy density K and the volume density Π of potential energy,

$$\mathcal{L}_i = \rho_i (\partial \mathbf{U}_i / \partial t)^2 / 2 - \Pi_i (I_k),$$

where $I_k = I_k(\mathbf{C})$ are the invariants of the Cauchy-Green deformation tensor $\mathbf{C}$ (see (2.4-5)). The strain energy density Π_i has the form of power series expansion

$$\Pi_i = (\lambda_i + 2\mu_i) I_1^2 / 2 - 2\mu_i I_2 + (l_i + 2m_i) I_1^3 / 3 - 2m_i I_1 I_2 + n_i I_3.$$

To derive the formal statement of the wave propagation problem in Lagrangian coordinates we use the Hamilton principle. For both longitudinal and shear waves the leading nonlinearity is square with respect to the *displacements*. Then, aiming to an asymptotic statement of the problem, we omit the nonlinear terms, which are cubic with respect to elastic strains, and we get the following equations:

$$
\begin{aligned}
\rho_i V_{i,tt} \quad & -\mu_i V_{i,zz} - (\lambda_i + 2\mu_i) V_{i,yy} - (\lambda_i + \mu_i) W_{i,yz} - \\
& \mu_i p_{1i}[(V_{i,z}^2)_y + (W_{i,y}^2)_y + 2(V_{i,y}V_{i,z})_z + 2(V_{i,z}W_{i,z})_z] - \mu_i p_{2i}(V_{i,y}^2)_y - \\
& \mu_i p_{3i}[(W_{i,z}^2)_y + 2(W_{i,z}V_{i,y})_y] - \mu_i p_{4i}[(W_{i,y}W_{i,z})_z + (V_{i,z}W_{i,y})_y + (V_{i,y}W_{i,y})_z] \\
= \quad & 0, \quad\quad\quad\quad\quad\quad\quad\quad\quad\quad\quad\quad\quad\quad\quad\quad\quad\quad\quad\quad\quad\quad\quad\quad (5.74)
\end{aligned}
$$

where the coefficients are defined as

$$
\begin{aligned}
p_{1i} &= (\lambda_i/2 + \mu_i + m_i/2)/\mu_i, \; p_{2i} = (3\lambda_i/2 + 3\mu_i + l_i + 2m_i)/\mu_i, \\
p_{3i} &= (\lambda_i/2 + l_i)/\mu_i, \; p_{4i} = (\mu_i + m_i)/\mu_i.
\end{aligned}
$$

The equation for the component $W_i(y, z, t)$ follows from (5.74) after formal replacements $V_i \rightarrow W_i$, $W_i \rightarrow V_i$, $y \rightarrow z$, $z \rightarrow y$ and it will not be written here for brevity.

To obtain the boundary conditions we introduce the following notations for the components of the stress tensor $\mathbf{T}_i$:

$$
\begin{aligned}
T_{i,yy} \quad & = \quad (\lambda_i + 2\mu_i)V_{i,y} + \lambda_i W_{i,z} + \\
& \mu_i p_{1i}(V_{i,z}^2 + W_{i,y}^2) + \mu_i p_{2i}V_{i,y}^2 + \mu_i p_{3i}(W_{i,z}^2 + 2W_{i,z}V_{i,y}) + \mu_i p_{4i}V_{i,z}W_{i,y}, \\
T_{i,yz} \quad & = \quad \mu_i(W_{i,y} + V_{i,z}) + \\
& 2\mu_i p_{1i}(W_{i,z}W_{i,y} + V_{i,y}W_{i,y}) + \mu_i p_{4i}(V_{i,z}W_{i,z} + V_{i,z}V_{i,y}).
\end{aligned}
$$

Then the boundary conditions at the free layer surface $y = h$ in each moment take the form:

$$
T_{1,yy} = 0, \; T_{1,yz} = 0, \quad\quad\quad\quad\quad\quad\quad\quad\quad\quad (5.75)
$$

At the interface $y = 0$ we shall study two models of elastic contact.

For the *first* one the boundary conditions are assumed to be the following:

$$
T_{1,yy} = T_{2,yy}, \; T_{1,yz} = 0, \; T_{2,yz} = 0, \; V_1 = V_2, \quad\quad\quad (5.76)
$$

while tangential components W_i of the displacement vector are supposed to be free, i.e., the *slippage* may occur at the interface.

The *second* model considered, namely, the *full contact*, requires the boundary conditions of continuity for both the displacement and stress components:

$$
T_{1,yy} = T_{2,yy}, \; T_{1,yz} = T_{2,yz}, \; V_1 = V_2, \; W_1 = W_2 \quad\quad\quad (5.77)
$$

For both models we have to assume evidently that:

$$
V_2, W_2 \rightarrow 0, \text{ for } y \rightarrow -\infty \quad\quad\quad\quad\quad\quad\quad (5.78)
$$

5.3.3 The layer and the half-space contact with slippage

Let us consider the problem (5.74)-(5.76) and (5.78), describing the layer and the half-space interaction established only by means of the normal stresses and displacements components. We introduce the dimensionless variables in the form: $\bar{z} = z/L$, $\bar{y} = y/H$, $Y = y/L$, $\bar{t} = c_{\tau 1}t/L$, $\bar{V}_i = V_i/U$, $\bar{W}_i = W_i/U$. Considering the propagation of sufficiently *long waves in a thin layer*, we introduce a small parameter as $\epsilon = H/L = U/L$ and suppose that dimensionless displacements depend on a phase variable $\theta = \bar{z} - c_0\bar{t}$ and a slow variable $X = \epsilon\bar{z}$ in both the layer and the half-space, whereas they depend on $\bar{y}$ in the layer, but on Y in the half-space.

We also introduce the following relations $a_i = c_{li}^2/c_{\tau i}^2$, $b_i = c_{\tau 1}^2/c_{\tau i}^2$, where $c_{li}, c_{\tau i}$ are the linear longitudinal and shear volume wave velocities respectively. Finally we get the following basic equations for waves in the layer, written in dimensionless form:

$$
\begin{aligned}
a_1 V_{1,yy} &+ p_{11}(W_{1,y}^2)_y + p_{21}(V_{1,y}^2)_y = \\
&-\epsilon\left[(a_1-1)W_{1,y\theta} + 2p_{31}(W_{1,\theta}V_{1,y})_y + p_{41}\left(V_{1,y}W_{1,y})_\theta + (V_{1,\theta}W_{1,y})_y\right)\right] \\
&-\epsilon^2\left[(a_1-1)W_{1,yX} + (1-b_1c_0^2)V_{1,\theta\theta} + p_{11}\left((V_{1,\theta}^2)_y + 2(V_{1,\theta}V_{1,y})_\theta\right) + \right. \\
&p_{31}(W_{1,\theta}^2)_y + 2(V_{1,y}W_{1,X})_y + p_{41}\left((V_{1,X}W_{1,y})_y + (W_{1,\theta}W_{1,y})_\theta + (V_{1,y}W_{1,y})_X\right)\right] \\
&-\epsilon^3\left[2p_{11}\left((V_{1,\theta}V_{1,X})_y + (V_{1,y}V_{1,X} + V_{1,\theta}W_{1,\theta})_\theta + (V_{1,y}V_{1,\theta})_X\right)\right. \\
&+2p_{31}(W_{1,\theta}W_{1,X})_y + 2(1-b_1c_0^2)V_{1,\theta X} + p_{41}\left((W_{1,\theta}W_{1,y})_X + (W_{1,y}W_{1,X})_\theta\right)\right] \\
&+O(\epsilon^4), \quad\quad (5.79)
\end{aligned}
$$

$$
\begin{aligned}
W_{1,yy} &+ 2p_{11}(V_{1,y}W_{1,y})_y = \\
&-\epsilon\left[(a_1-1)V_{1,y\theta} + p_{11}\left((W_{1,y}^2)_\theta + 2(W_{1,\theta}W_{1,y})_y\right) + p_{31}(V_{1,y}^2)_\theta + p_{41}(V_{1,y}V_{1,\theta})_y\right] \\
&-\epsilon^2\left[(a_1-1)V_{1,yX} + (a_1-b_1c_0^2)W_{1,\theta\theta} + p_{11}\left((W_{1,y}^2)_X + 2(W_{1,X}W_{1,y})_y\right) + \right. \\
&p_{31}(V_{1,y}^2)_X + 2(V_{1,y}W_{1,\theta})_\theta + p_{41}\left((V_{1,X}V_{1,y})_y + (V_{1,\theta}W_{1,y})_\theta + (V_{1,\theta}W_{1,\theta})_y\right)\right] - \\
&\epsilon^3\left[\left(p_{11}V_{1,\theta}^2 + p_{21}W_{1,\theta}^2\right)_\theta + 2p_{31}\left((V_{1,y}W_{1,X})_\theta + (V_{1,y}W_{1,\theta})_X\right)\right. \\
&+p_{41}\left((V_{1,X}W_{1,y})_\theta + (V_{1,\theta}W_{1,y})_X + (W_{1,\theta}V_{1,X})_y + (V_{1,\theta}W_{1,X})_y\right)\right] \\
&+2\epsilon^3(a_1-b_1c_0^2)W_{1,\theta X} + O(\epsilon^4). \quad\quad (5.80)
\end{aligned}
$$

where the upper bars were omitted for simplicity. It follows from (5.74) that nonlinear equations for waves in the half-space may be written similar to the equations (5.79), (5.80) but it turns out that in the *slippage contact* problem only the linear part of them will be of use to obtain the governing nonlinear equation for longitudinal strain waves:

$$
a_2 V_{2,YY} + (a_2-1)W_{2,Y\theta} + (1-b_2c_0^2)V_{2,\theta\theta} + 0(\epsilon) = 0, \quad\quad (5.81)
$$

$$
W_{2,YY} + (a_2-1)V_{2,Y\theta} + (a_2-b_2c_0^2)W_{2,\theta\theta} + 0(\epsilon) = 0, \quad\quad (5.82)
$$

And this is the first result following from the rigorous statement of the problem - slippage does not require the consideration of the nonlinearity of the substrate in our case!

The normal and tangential stresses components vanish at the free upper layer surface $y = 1$, that leads to the equations:

$$a_1 V_{1,y} + p_{11} W_{1,y}^2 + p_{21} V_{1,y}^2 = -\epsilon[(a_1 - 2)W_{1,\theta} + 2p_{31} V_{1,y} W_{1,\theta} + p_{41} V_{1,\theta} W_{1,y}]$$
$$-\epsilon^2[(a_1 - 2)W_{1,X} + p_{11} V_{1,\theta}^2 + p_{31}(W_{1,\theta}^2 + 2V_{1,y} W_{1,X}) + p_{41} V_{1,X} W_{1,y}]$$
$$-\epsilon^3[2p_{11} V_{1,\theta} V_{1,X} + 2p_{31} W_{1,\theta} W_{1,X}] + O(\epsilon^4), \tag{5.83}$$

$$W_{1,y} + 2p_{11} V_{1,y} W_{1,y} = -\epsilon[V_{1,\theta} + 2p_{11} W_{1,y} W_{1,\theta} + p_{41} V_{1,y} V_{1,\theta}]$$
$$-\epsilon^2[V_{1,X} + 2p_{11} W_{1,y} W_{1,X} + p_{41}(V_{1,y} V_{1,X} + V_{1,\theta} W_{1,\theta})] -$$
$$\epsilon^3 p_{41}(V_{1,X} W_{1,\theta} + V_{1,\theta} W_{1,X}) + O(\epsilon^4). \tag{5.84}$$

At the interface $y = 0$ $(Y = 0)$ we have to admit the continuity conditions for both the displacements:

$$V_1 = V_2 \tag{5.85}$$

and the stress normal components:

$$\mu_1 \left(a_1 V_{1,y} + p_{11} W_{1,y}^2 + p_{21} V_{1,y}^2 + \epsilon[(a_1 - 2)W_{1,\theta} + 2p_{31} V_{1,y} W_{1,\theta} + p_{41} V_{1,\theta} W_{1,y}] \right.$$
$$+\epsilon^2[(a_1 - 2)W_{1,X} + p_{11} V_{1,\theta}^2 + p_{31}(W_{1,\theta}^2 + 2V_{1,y} W_{1,X}) + p_{41} V_{1,X} W_{1,y}]$$
$$+\epsilon^3[2p_{11} V_{1,\theta} V_{1,X} + 2p_{31} W_{1,\theta} W_{1,X}]) =$$
$$\mu_2 \epsilon[a_2 V_{2,Y} + (a_2 - 2)W_{2,\theta}] + \epsilon^2 \left[(a_2 - 2)W_{2,X} + p_{12}(W_{2,Y}^2 + V_{2,\theta}^2) + p_{22} V_{2,Y}^2 + \right.$$
$$p_{32}(W_{2,\theta}^2 + 2V_{2,Y} W_{2,\theta}) + p_{42} V_{2,\theta} W_{2,Y} +$$
$$\epsilon^3[2p_{12} V_{2,\theta} V_{2,X} + 2p_{32}(W_{2,\theta} W_{2,X} + V_{2,Y} W_{2,X}) + p_{42} V_{2,X} W_{2,Y}]) + O(\epsilon^4), \tag{5.86}$$

while the tangential stress components vanish, i.e., boundary conditions are similar to the conditions (5.84). For the half-space at the interface $Y = 0$ it will be sufficient to use only the linear part of tangential stress components written in the form:

$$W_{2,Y} + V_{2,\theta} + O(\epsilon) = 0, \tag{5.87}$$

We shall find the solution of the problem (5.79)-(5.87), expanding displacements V_i, W_i in power series with respect to ϵ.

$$V_i = V_{0i} + \epsilon V_{1i} + \epsilon^2 V_{2i} + \ldots, \quad W_i = W_{0i} + \epsilon W_{1i} + \epsilon^2 W_{2i} + \ldots$$

Then in the leading order we obtain

$$V_{01} = 0, W_{01} = W(\theta, X), V_{02} = 0, W_{02} = 0.$$

In order ϵ we get the solution of the form:

$$V_{11} = \frac{2 - a_1}{a_1} y W_\theta, \quad W_{11} = 0, \quad V_{12} = 0, \quad W_{12} = 0.$$

In order ϵ^2 we obtain the solution

$$V_{21} = \frac{2 - a_1}{a_1} W_\theta + \left(\frac{a_1 - 4}{a_1^2} p_{31} - \frac{(a_1 - 2)^2}{a_1^3} p_{21} \right) (y - 1) W_\theta^2;$$

$$W_{21} = \left(\frac{2 - 3a_1}{a_1} + b_1 c_0^2 \right) \frac{y^2}{2} W_{\theta\theta},$$

that satisfies the boundary conditions (5.83), (5.84), (5.86) under the additional requirement:

$$c_0^2 = \frac{4(\lambda_1 + \mu_1)}{b_1(\lambda_1 + 2\mu_1)}, \tag{5.88}$$

and we *determined* the dimensionless value of the phase velocity via elasticity parameters and the linear tangential velocities ratio.

In the *half-space* the unknown functions V_{22}, W_{22} are to be found from the following linear problem:

$$a_2 V_{22,YY} + (a_2 - 1)W_{22,Y\theta} + (1 - b_2 c_0^2)V_{22,\theta\theta} = 0, \tag{5.89}$$

$$W_{22,YY} + (a_2 - 1)V_{22,Y\theta} + (a_2 - b_2 c_0^2)W_{22,\theta\theta} = 0, \tag{5.90}$$

subjected to the boundary conditions at $Y = 0$:

$$V_{22} = \frac{a_1 - 2}{a_1} W_X - \left(\frac{a_1 - 4}{a_1^2} p_{31} - \frac{(a_1 - 2)^2}{a_1^3} p_{21} \right) W_\theta^2, \tag{5.91}$$

$$W_{22,Y} + V_{22,\theta} = 0. \tag{5.92}$$

As it follows from (5.78) at $Y \to -\infty$

$$V_2, W_2 \to 0. \tag{5.93}$$

The solution of (5.89)-(5.93) may be obtained, using the Fourier transform method and, vanishing at $Y \to -\infty$, it may exist under additional condition of the form $1 - b_2 c_0^2 > 0$.

This condition *proves* that the wave phase velocity should be less than the bulk shear wave velocity in the half-space. It can be rewritten, using the c_0 definition in (5.88), as

$$c_{\tau 2} > 2c_{\tau 1}\sqrt{1 - \frac{c_{\tau 1}^2}{c_{l1}^2}}, \tag{5.94}$$

This relationship separates those elastic materials for the layer and the half-space respectively, for which the wave solution to (5.79)-(5.87) may really exist.

Some pairs of materials commonly used in acoustic experiments and devices (see Oliner, 1978) are presented in Table 5.4. Each column corresponds to the possible layer material and each row to the half-space material; a dash indicates that (5.94) is not valid for a pair.

When the condition (5.94) is valid, the relationships for V_{22}, W_{22} may be obtained using the standard procedure.

However, the most important result consists in the relationship for $a_2 V_{22,Y} + (a_2 - 2)W_{22,\theta}$ at $Y = 0$, and we write only it here due to the lack of space:

$$a_2 V_{22,Y} + (a_2 - 2)W_{22,\theta} \mid_{Y=0} = \left[\frac{(2 - b_2 c_0^2)^2 - 4\sqrt{(1 - b_2 c_0^2)(1 - b_2 c_0^2/a_2)}}{b_2 c_0^2 \sqrt{1 - b_2 c_0^2/a_2}} \right.$$
$$\times \mathcal{H}\left[\frac{a_1 - 2}{a_1}W_{\theta X} - \left(\frac{a_1 - 4}{a_1^2}p_{31} - \frac{(a_1 - 2)^2}{a_1^3}p_{21} \right)(W_\theta^2)_\theta \right]$$
$$(5.95)$$

Substituting (5.95) into the boundary condition (5.83) and solving the next order problem, $O(\epsilon^3)$, one can get the solvability condition in the form of the nonlinear integro-differential Benjamin-Ono equation written with respect to *longitudinal strain* $u = W_\theta$:

$$u_X + Bu_\theta^2 + C\mathcal{H}(u_{\theta\theta}) = 0. \qquad (5.96)$$

Table 5.4. Validity of condition for longitudinal strain wave existence for various layer - substrate (half-space) pairs.

Layer of: -on following Substrates:	Al	Mg	Mo	W	Cu	Pb	Sn	Brass
$LiNbO_3$	-	-	-	-	-	Yes	Yes	-
SiO_2 (Quartz)	-	-	-	-	-	Yes	Yes	Yes
GaAs	-	-	-	-	-	Yes	-	-
ZnO	-	-	-	-	-	Yes	-	-
C (Diamond)	Yes	Yes	Yes	Yes	Yes	Yes	Yes	Yes
MgO	Yes	Yes	Yes	Yes	Yes	Yes	Yes	Yes

In (5.95,5.96) $\mathcal{H}$ is the Hilbert transform and the coefficients are defined as follows:

$$B = \frac{p_{21}(a_1 - 2)^2 - a_1 p_{31}(a_1 - 4)}{a_1^2(a_1 - 2)},$$

$$C = \frac{\mu_1 b_2 c_0^2 \sqrt{1 - b_2 c_0^2/a_2}(a_1 - 1)(a_1 b_1 c_0^2 - a_1 - 2)}{2\mu_2(a_1 - 2)[(2 - b_2 c_0^2)^2 - 4\sqrt{(1 - b_2 c_0^2)(1 - b_2 c_0^2/a_2)}]}.$$

The *periodical wave* solution to the equation (5.96), well known after Benjamin and Ono, can be written in the form

$$u = \frac{C|\gamma|\sqrt{1 - \beta^2}}{B(1 - \beta\cos(\gamma(\theta - sX)))} - \frac{C|\gamma|}{B}, \qquad (5.97)$$

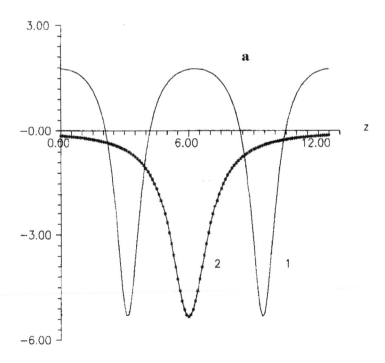

Figure 5.11: Periodical (1) and solitary (2) wave base-band solutions to the Benjamin-Ono equation, calculated for the brass-quartz pair.

where $s = C \mid \gamma \mid (1/\sqrt{1-\beta^2} - 2)$, and β and γ are free parameters. The *solitary wave* solution has the form of the algebraic soliton:

$$u = \frac{2C \mid \beta \mid}{B((\theta - sX)^2 + \beta^2)}, \qquad (5.98)$$

where $s = C/|\beta|$, and β is free. Shapes of the wave solutions (5.97), (5.98) depend on the $\mathrm{sgn}C/B$. It may be positive or negative, depending on layer and the half-space elasticity.

For example, $C/B = -2.66$ for the brass layer on quartz substrate. The corresponding periodical and solitary waves profiles are shown in Figure 5.11. On the contrary, the positive value $C/B = 1.08$ is valid for the brass layer imposed on MgO, and the corresponding wave profiles shown in Figure 5.12 are different.

Let us examine now the existence of the *envelope non-linear wave* solution in the slippage contact problem.

Following the standard method (see, e.g., Newell, 1985), we introduce formally a small parameter $\delta \ll 1$ and suppose that the function W depends on both fast variable θ and the slow variables $Z_1 = \delta X$, $Z_2 = \delta \theta$ and $\tau = \delta^2 X$. However it would lead to considerable formal difficulties in introducing the slow variable Z_2 in the Hilbert transform in equation (5.96). That is why the more suitable way is to derive the resulting envelope waves equations starting with the equations (5.89)-(5.93). Introducing both the fast and slow variables, we will find the solution of the

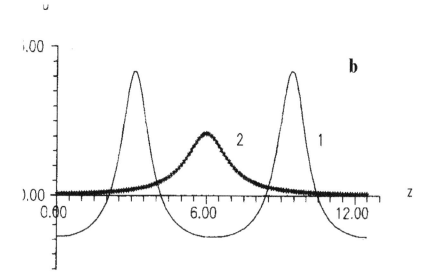

Figure 5.12: The same is true for a brass layer on the MgO substrate, but the pulse polarity is different.

form

$$V_{22} = \delta V_{220} + \delta^2 V_{221} + \dots,$$
$$W_{22} = \delta W_{220} + \delta^2 W_{221} + \dots,$$
$$W = \delta W_0 + \delta^2 W_1 + \dots. \tag{5.99}$$

Substituting these series into (5.89)-(5.93) and solving problems in each order on δ, using the Fourier transform method, one can find in order δ the solvability condition in the form of the non-linear equation for the strain function $u = W_{0,\theta}$:

$$u_X + B\,\mathcal{H}(u_{\theta\theta}) = 0.$$

Hence the solution is:

$$u = \alpha(\tau,\ Z_1,\ Z_2)\exp(i\,m(\theta + C\ |\ m\ |\ X)) + (^*), \tag{5.100}$$

where $(^*)$ is used for a complex conjugate term. Next order solution results in the solvability condition of the form:

$$u_{1,X} + u_{Z_1} + B(u^2)_\theta + C\mathcal{H}(u_{1,\theta\theta}) + D\mathcal{H}(u_{\theta Z_2}) = 0, \tag{5.101}$$

where $u_1 = W_{1,\theta}$ and $D = D(a_i, b_i, \mu_i)$. From equation (5.101) we conclude that $\alpha = \alpha(\tau,\ \xi),\ \xi = Z_2 + D\ |\ m\ |\ Z_1$ and

$$u_1 = \frac{\alpha^2 B}{|\ m\ |\ C}\exp 2i\,m(\theta + C\ |\ m\ |\ X) + (^*) + \gamma(\tau,\ \xi). \tag{5.102}$$

In the next order problem the nonlinear Schrödinger (NLS) equation arises from the solvability condition:

$$i\,\mathrm{sgn}(m)\,\alpha_T + \frac{2B^2(2C - D)}{CD}\alpha^2\alpha^* - E\alpha_{\xi\xi} = 0, \tag{5.103}$$

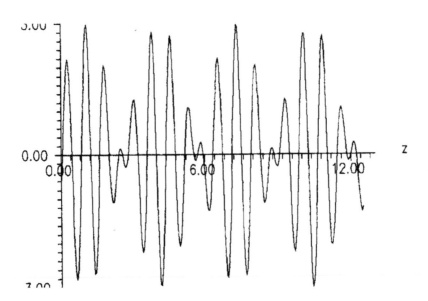

Figure 5.13: An envelope periodic wave solution to the NLS equation.

where E is expressed in terms of a_i, b_i, μ_i also.

The relationships for D and E are very cumbersome and not written here.

The product of the nonlinear coefficient times the dispersion one in the equation (5.103) may be either negative or positive, depending on elasticity of materials chosen. The former case corresponds to the modulation instability while the latter case results in non-increasing disturbances of the initial harmonic wave. Hence in the former case, $E(2C - D)/CD < 0$, we can obtain the solution for the strain wave u in the form, depending on the periodical elliptic Jacobi functions (see, e.g., Newell, 1985):

$$u = \sqrt{\frac{CDE}{4B^2(D - 2C)}}\, \beta \, \mathrm{dn}(\beta(\xi + 2k\,\mathrm{sgn}(m)\,E\,\tau),\,\kappa)\exp\left[i\,m(\theta + C\mid m\mid X)+ \right.$$
$$\left. ik\xi - i\,\mathrm{sgn}(m)\,((2 - \kappa^2)\beta^2 - k^2)\,E\tau\right] \tag{5.104}$$

The typical envelope periodic wave form is shown in Figure 5.13. When $E(2C - D)/CD > 0$, the equation (5.100) has the exact solution of the form:

$$u = R\exp\left(i\,m(\theta + C\mid m\mid X + i\varphi(\xi,\tau) + i\,s\,\xi/2 - i(q + s^2/4)\mathrm{sgn}(m)\,E\,\tau) + (^{*}),\right. \tag{5.105}$$

where[1]

$$R^2 = \gamma^2 - \frac{CDE\kappa^2\beta^2}{B^2(2C - D)}\mathrm{cn}^2[\beta(\xi + sE\mathrm{sgn}(m)\tau),\kappa],$$

$$\varphi_\xi = Q/R^2, \quad Q^2 = \frac{B^2(2C - D)\gamma^6}{CED} - (2\kappa^2 - 1)\gamma^4\beta^2 - \frac{CDE\gamma^2\beta^4\kappa^2(1 - \kappa^2)}{B^2(2C - D)},$$

$$q = \frac{3B^2(2C - D)\gamma^2}{CDE} - (2\kappa^2 - 1)\beta^2.$$

---

[1]Misprints were corrected in these formuli.

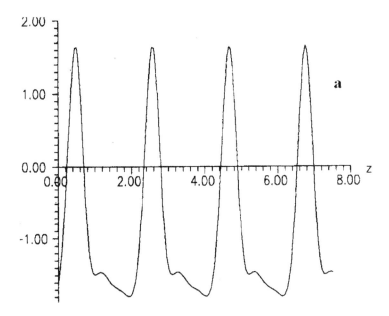

Figure 5.14: A typical wave evolution for the NLS equation solution (a)-(c) is shown in different time moments: $t_a < t_b < t_c$.

In the limit $\kappa \to 1$ the solution (5.105) coincides with the one already obtained by Newell (1985). The typical wave form strongly depend on Q, γ, β and m values. In Figures 5.14, 5.15, 5.16 typical wave evolution is shown for $Q > m$, and the magnitude of R is assumed not to vanish. In another special case, corresponding to the small R magnitude and $Q < m$, one could see only slow and small disturbances on the harmonic wave shape.

The appearance of either B-O travelling waves (5.96), (5.97) or envelope travelling wave solutions (5.104), (5.105) is defined by the initial condition for the longitudinal strain u. We shall discuss it in detail below, comparing our results with recent experiments mentioned in (Ewen et al. (1982), Cho and Miyagawa, (1993), Dyakonov et al. (1993), Nayanov (1986)).

5.3.4 The full contact of the layer and the half-space

Now we shall consider the *full contact* problem, in which all components of both the displacements and the stresses are continuous at the layer–half-space interface. We shall use the same scales and the small parameter as in the previous section, but now it is more convenient to introduce a time-dependent slow variable $T = \epsilon t$ instead of X.

The terms of the third order in ϵ in the layer equations and of the second order in ϵ in the half-space equations can be omitted because they will not appear, when the resulting (asymptotic) nonlinear equation will be derived. After some transformations we obtain the equations for *displacements in a layer* in the form:

$$a_1 V_{1,yy} \quad + p_{11}(W_{1,y}^2)_y + p_{21}(V_{1,y}^2)_y =$$

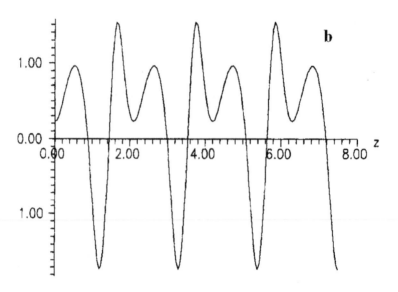

Figure 5.15: The same solution for $t_b > t_a$.

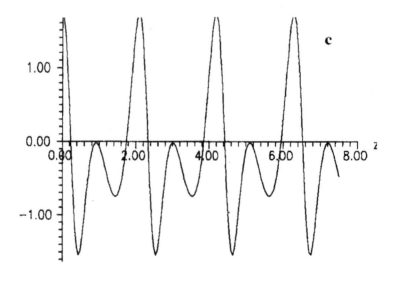

Figure 5.16: The same wave in $t_c > t_b > t_a$.

$$-\epsilon\left[(a_1-1)W_{1,y\theta}+2p_{31}(W_{1,\theta}V_{1,y})_y+p_{41}\left((V_{1,y}W_{1,y})_\theta+(V_{1,\theta}W_{1,y})_y\right)\right]$$
$$-\epsilon^2\left[(1-b_1c_0^2)V_{1,\theta\theta}+p_{11}\left((V_{1,\theta}^2)_y+2(V_{1,\theta}V_{1,y})_\theta\right)+p_{31}(W_{1,\theta}^2)_y+\right.$$
$$\left.p_{41}(W_{1,\theta}W_{1,y})_\theta\right]+O(\epsilon^3),\tag{5.106}$$

$$W_{1,yy}+2p_{11}(V_{1,y}W_{1,y})_y=$$
$$-\epsilon\left[(a_1-1)V_{1,y\theta}+p_{11}\left((W_{1,y}^2)_\theta+2(W_{1,\theta}W_{1,y})_y\right)+\right.$$
$$\left.p_{31}(V_{1,y}^2)_\theta+p_{41}(V_{1,y}V_{1,\theta})_y\right]$$
$$-\epsilon^2\left[(a_1-b_1c_0^2)W_{1,\theta\theta}+2p_{31}(V_{1,y}W_{1,\theta})_\theta+p_{41}\left((V_{1,\theta}W_{1,y})_\theta+(V_{1,\theta}W_{1,\theta})_y\right)\right]$$
$$+O(\epsilon^3).\tag{5.107}$$

The following equations are valid for *displacements in a half-space:*

$$a_2V_{2,YY}+(a_2-1)W_{2,Y\theta}+(1-b_2c_0^2)V_{2,\theta\theta}=$$
$$-\epsilon\left[2b_2c_0V_{2,T\theta}+p_{12}\left((V_{2,\theta}^2)_Y+(W_{2,Y}^2)_Y+2(V_{2,Y}V_{2,\theta})_\theta+2(V_{2,\theta}W_{2,\theta})_\theta\right)\right.$$
$$+p_{22}(V_{2,Y}^2)_Y+p_{32}\left((W_{2,\theta}^2)_Y+2(V_{2,Y}W_{2,\theta})_Y\right)+$$
$$\left.p_{42}\left((W_{2,Y}W_{2,\theta})_\theta+(V_{2,\theta}W_{2,Y})_Y+(V_{2,Y}W_{2,Y})_\theta\right)\right]+O(\epsilon^2);\tag{5.108}$$

$$W_{2,YY}+(a_2-1)V_{2,Y\theta}+(a_2-b_2c_0^2)W_{2,\theta\theta}=$$
$$-\epsilon\left[2b_2c_0W_{2,T\theta}+p_{12}\left((V_{2,\theta}^2)_\theta+(W_{2,Y}^2)_\theta+2(W_{2,Y}W_{2,\theta})_Y+2(V_{2,Y}V_{2,Y})_Y\right)\right.$$
$$+p_{22}(W_{2,\theta}^2)_\theta+p_{32}\left((V_{2,Y}^2)_\theta+2(V_{2,Y}W_{2,\theta})_\theta\right)+$$
$$\left.p_{42}\left((W_{2,Y}V_{2,\theta})_\theta+(V_{2,\theta}V_{2,Y})_Y+(V_{2,\theta}W_{2,\theta})_Y\right)\right]+O(\epsilon^2).\tag{5.109}$$

The boundary conditions at the upper layer surface have the form similar to the equations (5.83), (5.84):

$$a_1V_{1,y}+p_{11}W_{1,y}^2+p_{21}V_{1,y}^2=-\epsilon\left[(a_1-2)W_{1,\theta}+2p_{31}V_{1,y}W_{1,\theta}+p_{41}V_{1,\theta}W_{1,y}\right]-$$
$$\epsilon^2\left[p_{11}V_{1,\theta}^2+p_{31}W_{1,\theta}^2\right]+O(\epsilon^3),\tag{5.110}$$

$$W_{1,y}+2p_{11}V_{1,y}W_{1,y}=-\epsilon\left[V_{1,\theta}+2p_{11}W_{1,y}W_{1,\theta}+p_{41}V_{1,y}V_{1,\theta}\right]-\epsilon^2p_{41}V_{1,\theta}W_{1,\theta}+O(\epsilon^3).\tag{5.111}$$

However, the boundary conditions at the interface $y=0$ ($Y=0$) are now of different form:

$$V_1=V_2,\quad W_1=W_2,\tag{5.112}$$

$$\mu_1\left(a_1V_{1,y}+p_{11}W_{1,y}^2+p_{21}V_{1,y}^2+\epsilon\left[(a_1-2)W_{1,\theta}+2p_{31}V_{1,y}W_{1,\theta}+p_{41}V_{1,\theta}W_{1,y}\right]\right.$$
$$\left.+\epsilon^2\left[p_{11}V_{1,\theta}^2+p_{31}(W_{1,\theta}^2)\right]\right)=$$
$$\mu_2\left(\epsilon\left[a_2V_{2,Y}+(a_2-2)W_{2,\theta}\right]+\epsilon^2\left[p_{12}(W_{2,Y}^2+V_{2,\theta}^2)+\right.\right.$$
$$\left.\left.p_{22}V_{2,Y}^2+p_{32}(W_{2,\theta}^2+2V_{2,Y}W_{2,\theta})+p_{42}V_{2,\theta}W_{2,Y}\right]\right)+O(\epsilon^3).\tag{5.113}$$

$$\mu_1\left(W_{1,y}+2p_{11}V_{1,y}W_{1,y}+\epsilon\left[V_{1,\theta}+2p_{11}W_{1,y}W_{1,\theta}+p_{41}V_{1,y}V_{1,\theta}\right]+\epsilon^2p_{41}V_{1,\theta}W_{1,\theta}\right)=$$
$$\mu_2\left(\epsilon\left[W_{2,Y}+V_{2,\theta}\right]+\epsilon^2\left[2p_{12}(W_{2,y}W_{2,\theta}+V_{2,Y}W_{2,Y})+p_{42}(V_{2,y}V_{2,\theta}+V_{2,\theta}W_{2,\theta})\right]\right)$$
$$+O(\epsilon^3).\tag{5.114}$$

Moreover, at $Y \to -\infty$ displacements should vanish:

$$V_2, \ W_2 \ \to \ 0. \tag{5.115}$$

We will find a solution of the problem (5.106)-(5.115) in the form of power series expansions:

$$V_i \ = \ V_{0i} + \epsilon V_{1i} + \epsilon^2 V_{2i} + \dots, \quad W_i \ = \ W_{0i} + \epsilon W_{1i} + \epsilon^2 W_{2i} + \dots$$

To obtain an asymptotic solution of the problem in each order of ϵ we will find by the simplest integration the corresponding displacement functions in the layer V_{j1}, W_{j1}, using only the boundary conditions (5.110), (5.111) at the upper surface. Substituting these functions to the boundary conditions (5.112), we obtain the displacements in the half-space, solving by means of the Fourier transform method the equations in corresponding order subjected to the boundary conditions (5.112) and (5.115). Finally, using the values of V_{ji}, W_{ji} in the boundary conditions (5.113), (5.114), we determine the solution parameters values and obtain an additional *solvability condition* in each order of the problem under consideration.

The transformations described above are cumbersome, and we write here only the most important results. Following this scheme we obtain in the leading order problem both *nonzero longitudinal and shear displacements in the layer*, expressed in terms of new unknown functions:

$$V_{01} \ = \ V_0(\theta, T), \quad W_{01} \ = \ W_0(\theta, T) \tag{5.116}$$

Introducing new functions ϕ_0 and ψ_0, such that

$$v_{02} \ = \ \phi_{0,Y} + \psi_{0,\theta}, \quad w_{02} \ = \ \phi_{0,\theta} - \psi_{0,Y}$$

we derive the following leading order problem in the half-space:

$$a_2 \phi_{0,YY} + (a_2 - b_2 c_0^2)\phi_{0,\theta\theta} \ = \ 0 \tag{5.117}$$

$$\psi_{0,YY} + (1 - b_2 c_0^2)\psi_{0,\theta\theta} \ = \ 0 \tag{5.118}$$

where at the interface $Y = 0$

$$\phi_{0,\theta} - \psi_{0,Y} \ = \ W_0, \quad \phi_{0,Y} + \psi_{0,\theta} \ = \ V_0 \tag{5.119}$$

while at infinity $Y \to -\infty$

$$\phi_{0,\theta} - \psi_{0,Y} \ \to \ 0, \quad \phi_{0,Y} + \psi_{0,\theta} \ \to \ 0 \tag{5.120}$$

The solution to the problem (5.117)-(5.120) can be obtained by means of the Fourier transform and allows us to establish the following relationships for dicplacement functions $V_{02}, W_{0,2}$:

$$V_{02} \ = \ \frac{1}{2\pi} \int_{-\infty}^{\infty} e^{ik\theta} dk \int_{-\infty}^{\infty} \left[\left(\frac{i \mid k \mid \sqrt{1 - b_2 c_0^2 / a_2}}{k} W_0 + \sqrt{(1 - b_2 c_0^2)(1 - b_2 c_0^2 / a_2)} V_0 \right) \right.$$

$$e^{\sqrt{1 - b_2 c_0^2 / a_2} \mid k \mid Y} - \left(V_0 + \frac{i \mid k \mid \sqrt{1 - b_2 c_0^2 / a_2}}{k} W_0 \right) e^{\sqrt{1 - b_2 c_0^2} \mid k \mid Y} \right]$$

$$\times e^{-ik\theta'} d\theta', \tag{5.121}$$

$$W_{02} = \frac{1}{2\pi} \int_{-\infty}^{\infty} e^{ik\theta} dk \int_{-\infty}^{\infty} \left[\left(\frac{i \mid k \mid \sqrt{1 - b_2 c_0^2}}{k} V_0 - W_0 \right) e^{\sqrt{1 - b_2 c_0^2/a_2} \mid k \mid Y} - \right.$$

$$\left. \left(\frac{i \mid k \mid \sqrt{1 - b_2 c_0^2}}{k} V_0 - \sqrt{(1 - b_2 c_0^2)(1 - b_2 c_0^2/a_2)} W_0 \right) e^{\sqrt{1 - b_2 c_0^2} \mid k \mid Y} \right]$$

$$\times e^{-ik\theta'} d\theta', \tag{5.122}$$

In the next order, namely, in $O(\epsilon^1)$, we obtain *in the layer*:

$$V_{11} = V_1(\theta, T) - \frac{a_1 - 2}{a_1} y\, W_{0,\theta},$$

$$W_{11} = W_1(\theta, T) - yV_{0,\theta}$$

The solution to *the half-space* problem is assumed to have the form:

$$V_{12} = \phi_{1,Y} + \psi_{1,\theta},$$

$$W_{12} = \phi_{1,\theta} - \psi_{1,Y}.$$

Therefore one can find from (5.108), (5.109) that the following equations are valid for the functions ϕ_1 :

$$\Delta_{\theta Y} \left(\psi_{1,YY} + (1 - b_2 c_0^2)\psi_{1,\theta\theta} \right) = -2c_0 b_2 \Delta_{\theta Y} \psi_{0,\theta T} - \mathbf{D_1} \tag{5.123}$$

and for ψ_1 :

$$\left(a_2 \phi_{1,YY} + (a_2 - b_2 c_0^2)\phi_{1,\theta\theta} \right)_\theta = \left[a_2 \psi_{1,YY} + (1 - b_2 c_0^2)\psi_{1,\theta\theta} \right]_Y$$
$$- 2b_2 c_0(\phi_{1,\theta} - \psi_{1,Y})_{\theta\theta} - \mathbf{D_2} \tag{5.124}$$

where the Laplasian is defined as $\Delta_{\theta Y} = \partial^2/\partial\theta^2 + \partial^2/\partial Y^2$, and the following differential expressions were introduced:

$$\mathbf{D_1} = 2p_{12} \left((V_{02,\theta}(V_{02,Y} + W_{02,\theta}))_{\theta\theta} - (W_{02,Y}(V_{02,Y} + W_{02,\theta}))_{YY} \right) +$$
$$(p_{22} - p_{32}) \left(v_{02,Y}^2 - w_{02,\theta}^2 \right)_{Y\theta}$$
$$+ p_{42} \left((w_{02,Y}(v_{02,Y} + w_{02,\theta}))_{\theta\theta} - (v_{02,\theta}(v_{02,Y} + w_{02,\theta}))_{YY} \right).$$

$$\mathbf{D_2} = p_{12} \left((v_{2,\theta}^2)_\theta + (w_{2,Y}^2)_\theta + 2(w_{2,Y}w_{2,\theta})_Y + 2(v_{2,Y}w_{2,Y})_Y \right)$$
$$+ p_{22}(w_{2,\theta}^2)_\theta + p_{32} \left((v_{2,Y}^2)_\theta + 2(v_{2,Y}w_{2,\theta})_\theta \right) +$$
$$p_{42} \left((w_{2,Y}v_{2,\theta})_\theta + (v_{2,\theta}v_{2,Y} + v_{2,\theta}w_{2,\theta})_Y \right)$$

Subsequently, we have to announce the main result, partially negative: due to the presence of these expressions - the solutions similar to the wave solutions (5.97), (5.98) of the B-O equation do not appear in the full contact problem. It will be proved formally below.

The boundary conditions are written at $Y = 0$ as:

$$\phi_{0,\theta} - \psi_{0,Y} = 0, \quad \phi_{0,Y} + \psi_{0,\theta} = 0, \tag{5.125}$$

while at $Y \rightarrow -\infty$

$$\phi_{0,\theta} - \psi_{0,Y} \rightarrow 0, \quad \phi_{0,Y} + \psi_{0,\theta} \rightarrow 0. \tag{5.126}$$

The solution to the problem (5.123)-(5.126) can be obtained using the Fourier transform method and the variation of constants method. Substitution of the relationships for the functions v_{11}, w_{11}, v_{02} and w_{02} into the boundary conditions (5.113), (5.114) gives the *solvability* conditions in form of the coupled equations for functions W_0 and V_0:

$$\left(b_2c_0^2 - 2 + 2\sqrt{(1 - b_2c_0^2(1 - b_2c_0^2/a_2))}\right) W_{0,\theta} - b_2c_0^2\sqrt{1 - b_2c_0^2}\mathcal{H}(V_{0,\theta}) = 0, \tag{5.127}$$

$$b_2c_0^2\sqrt{1 - b_2c_0^2/a_2}W_{0,\theta} + \left(2 - b_2c_0^2 - 2\sqrt{(1 - b_2c_0^2)(1 - b_2c_0^2/a_2)}\right)\mathcal{H}(V_{0,\theta}) = 0. \tag{5.128}$$

The determinant of (5.127), (5.128) is taken equal to zero, which yields the algebraic equation for the phase velocity c_0 of a wave:

$$(2 - b_2c_0^2)^2 - 4\sqrt{(1 - b_2c_0^2)(1 - b_2c_0^2/a_2)} = 0$$

Therefore $c_0 = c_R/c_{T1}$, where c_R is the Rayleigh linear wave velocity in the half-space, and respectively, the relationship for V and W follows from the equations (5.127), (5.128) in the form:

$$V_0 = \frac{2 - b_2c_0^2}{2\sqrt{1 - b_2c_0^2}}\mathcal{H}(W_0). \tag{5.129}$$

In the next order, $O(\epsilon^2)$, following the same procedure, one can obtain the solution for the functions V_{21}, W_{21}, V_{11} and W_{11}. Again, the forms for them are cumbersome, and we have to present here only the most important results. In particular, from the boundary conditions (5.113), (5.114) the equations arise, which define the unknown functions V_1, W_1, and are now the inhomogeneous ones in contrast to the equations (5.127), (5.128). One can show that the solution to these equations may exist under the *solvability* condition, having the form of a complicated nonlinear integro-differential equation for the *strain* function $u = W_{0,\theta}$:

$$Au_T + B\mathcal{H}(u_{\theta\theta}) + C(u)_\theta^2 + D(\mathcal{H}(u)^2)_\theta =$$

$$= E\frac{1}{2\pi}\int_{-\infty}^{\infty}\exp^{ik\theta}dk\int_{-\infty}^{0}\exp^{\sqrt{1-b_2c_0^2/a_2}|k|Y}dY\int_{-\infty}^{\infty}\mathbf{D_1}\exp^{-ik\theta'}d\theta'$$

$$+F\frac{1}{2\pi}\int_{-\infty}^{\infty}\exp^{ik\theta}dk\int_{-\infty}^{0}\exp^{\sqrt{1-b_2c_0^2/a_2}|k|Y}dY\int_{-\infty}^{\infty}\mathbf{D_{2Y}}\exp^{-ik\theta'}d\theta'$$

$$-G\frac{1}{2\pi}\int_{-\infty}^{\infty}\exp^{ik\theta}dk\int_{-\infty}^{0}\exp^{\sqrt{1-b_2c_0^2}|k|Y}dY\int_{-\infty}^{\infty}\mathbf{D_1}\exp^{-ik\theta'}d\theta'. \tag{5.130}$$

The coefficients $A - G$ here depend on both the layer and the half-space *linear* elastic moduli λ_i, μ_i, while only on the *nonlinear* elastic moduli l_2, m_2, n_2 of half-space's material, as follows:

$$A = \frac{4b_2c_0(4(1 + a_2 - 2b_2c_0^2) - a_2(2 - b_2c_0^2)^3)}{a_2(2 - b_2c_0^2)^2},$$

$$B = \frac{\mu_1 b_2 c_0^2}{\mu_2}\left(b_1 c_0^2\sqrt{1 - \frac{b_2 c_0^2}{a_2}} + \sqrt{1 - b_2 c_0^2}(b_1 c_0^2 - 4 + 4\frac{b_1 c_0^2}{a_1})\right),$$

$$C = \frac{(2 - b_2 c_0^2)b_2 c_0^2}{4a_2^2(a_2 - b_2)}\left(p_{12}[16(1 - a_2)^2 + 4a_2 b_2 c_0^2(2 - a_2)]\right.$$
$$+2p_{22}(2 - a_2)(a_2^2 + a_2 - b_2 c_0^2)+$$

$$p_{32}[(2 - a_2^2)^3 - a_2(a_2^2 - 8) + 2a_2 b_2 c_0^2(a_2 - 8)] + 2p_{42}[4(1 - a_2) + a_2 b_2 c_0^2]\right),$$

$$D = \left[4a_2(2 - b_2 c_0^2)(5 - b_2 c_0^2) - (a_2 - b_2 c_0^2)(a_2(2 - b_2 c_0^2)^2 + 4(1 - b_2 c_0^2))\right]$$
$$(b_2 c_0^2)(2p_{12} - p_{42})(2 - b_2 c_0^2)\left[8a_2(1 - b_2 c_0^2)(a_2 - b_2 c_0^2)\right]^{-1},$$

$$E = (2 - b_2 c_0^2)^2/2, F = \frac{(2 - b_2 c_0^2)^2 b_2 c_0^2}{2(a_2 - b_2 c_0^2)}, G = 2 - b_2 c_0^2$$

We can verify now that due to the integrals proportional to $\mathbf{D_1}, \mathbf{D_2}$ the equation (5.130) may possess exact solutions (5.97), (5.98), typical for the Benjamin-Ono equation only if $a_2 = 1$, that is $c_{l2} = c_{\tau 2}$, which is impossible from the physical view point.

The feasibility of another exact travelling wave solutions is restricted by the functions set for which the Hilbert transform is known. That is why we have not obtained any exact solutions of the equation (5.130); it seems to be quite probable that the long base-band strain waves solutions cannot exist in the system, when the full contact is considered.

However the envelope waves may appear, and their evolution is described again by the NLS equation. This equation may be derived using the procedure similar to the one used in the previous section for obtaining the equation (5.103). Then no new physically important solutions, except these envelope waves, may be found in the full contact problem, except those described by the expressions (5.104), (5.105) and having the shapes shown in figures.

5.3.5 On various mathematical models of nonlinear waves in a layered medium

We have studied above both the base-band and the envelope *long* nonlinear strain waves propagation in elastic system, consisting of the thin elastic layer superimposed on elastic half-space. We found that base-band waves may appear only when slippage comes into play at the layer–half-space interface.

Recently in several papers by Ewen et al. (1982), Cho and Miyagawa (1993), Nayanov (1986) the Korteveg-de Vries equation was considered as a model one to describe both longitudinal and SH shear strain base-band waves propagation in the layered half-space when the full contact problem is considered. Moreover the linear model analysis performed by Tiersten (1969) did not predict the linearized Benjamin-Ono equation appearance to describe the problem under study.

For this reason it seems to be instructive to discuss the linear analysis of the problem. Considering the solution (2.13) in (Tiersten, 1969), one can see that only positive values of wave number were taken into account there. When taken into account, the negative wave number values lead immediately to replacements of the terms $\exp(-ky\sqrt{1-c_{\tau1}^2 c_0^2/c_{l2}^2})$ and $\exp(-ky\sqrt{1-c_{\tau1}^2 c_0^2/c_{\tau2}^2})$ in (2.13) in (Tiersten, 1969) with the terms $\exp(-\mid k\mid y\sqrt{1-c_{\tau1}^2 c_0^2/c_{l2}^2})$ and $\exp(-\mid k\mid y\sqrt{1-c_{\tau1}^2 c_0^2/c_{\tau2}^2})$ respectively, newly written in our notations for convenience. Then instead of the dispersion relation (4.5) from (Tiersten, 1969) in thin layer assumption, we obtain the following one:

$$A_0 + A_1 \mid k \mid + A_2 k^2 = 0, \qquad (5.131)$$

where A_j are the same as in (Tiersten, 1969) and may be expressed in our notation in the form:

$$
\begin{aligned}
A_0 &= (2 - b_2 c^2)^2 - 4\sqrt{(1 - b_2 c^2)}(1 - b_2 c^2/a_2); \\
A_1 &= \frac{\mu_1 b_2 c^2}{\mu_2}\left(c_0^2\sqrt{1 - b_2 c^2/a_2} + \sqrt{1 - b_2 c^2}(c_0^2 - 4 + 4\frac{b_1}{a_1})\sqrt{1 - b_2 c^2}\right); \\
A_2 &= \frac{\mu_1^2 c^2}{\mu_2^2}\left(4 - c^2 - 4\frac{b_1}{a_1}\right)\left(1 - (1 - b_2 c^2)(1 - \frac{b_2 c^2}{a_2})\right).
\end{aligned}
\qquad (5.132)
$$

Finding the dimensionless phase velocity in the long waves limit $\mid k \mid\, <<\, 1$ as a power series in k

$$c = c_0 + \mid k \mid c_1 + k^2 c_2 + \dots , \qquad (5.133)$$

substituting it into (5.131) and using (5.132), we can prove that $c_0 = c_R/c_{\tau1}$ and

$$c_1 = \frac{\mu_1 a_2 c_0 (2 - b_2 c_0^2)^2\left(b_1 c_0^2\sqrt{1 - b_2 c_0^2/a_2} + \sqrt{1 - b_2 c_0^2}(b_1 c_0^2 - 4 + 4 b_1 c_0^2/a_1)\right)}{4\mu_2(4(1 + a_2 - b_2 c_0^2) + a_2(2 - b_2 c_0^2)^3)}$$

$$(5.134)$$

These results define the first two terms in (5.133) and lead to the dispersion relation typical for linearized Benjamin-Ono equation.

The values of c_0, c_1 coincide with those followed from the solution of linearized equation (5.130). One can see that the first and the third terms in (5.133) together give the linearized Korteveg-de Vries dispersion relation, however the third term is of much smaller order of magnitude in comparison with the second one, except for the layer material, having the Lamé coefficient μ_1 much greater than the half-space coefficient μ_2. The last case seems not to be realistic, because typical shear moduli ratio for the materials used in acoustic experiments does not exceed three. Therefore the KdV equation is unlikely to be valid to describe strain waves in layered half-space, and waves observed by Nayanov (1986) *cannot be the KdV solitons* as it was proposed.

The similar waves, reported in (Cho and Miyagawa, 1993), are also unlikely the KdV solitons because the grating wave guide structure used there could hardly result in such a critical qualitative differences in dispersion relation.

In contrast to it, we believe that the waves observed by Nayanov (1986) and Cho and Miyagawa (1993) may be the periodical waves solution (5.105) to the NLS equation (5.103). The evolution of such wave shown in Figures 5.14-5.16 is surprisingly similar to the wave evolution observed and reported by Nayanov[2]. It is evident when comparing the wave shapes shown in Figures 5.14-5.16 with Figures 5.17.

Note that we do not insist on the slippage appearance in those experiments because we have found that such a wave may exist in the full contact problem also. We cannot provide rigorous quantitative estimations due to lack of some important data in papers cited above.

It is to be noted that for SH (Love) waves we obtain the completely different dispersion relation in the long wave limit, namely,

$$
c = c_{\tau 2} \left(k - \frac{\mu_1^2 k^3}{2\mu_2^2} \frac{c_{\tau 2}^2}{c_{\tau 1}^2} - 1)^2 \right) + O(k^5), \tag{5.135}
$$

which is similar to the one obtained, e.g., by Maradudin (1987), even with negative wave numbers values being taken into account. The dispersion relation (5.135) corresponds to the linearized Korteveg-de Vries equation. Therefore the application of the KdV equation for nonlinear Love waves propagation in layered half-space, made by Sakuma and Nishiguchi (1990), has no contradiction with the linear waves analysis.

The reason of such a qualitatively different description of nonlinear longitudinal and SH strain waves in layered elastic half-space may be based on their different physical nature. The SH wave became the surface one (the Love wave) for the layered half-space only, while the longitudinal surface wave (the Rayleigh wave) may exist in absence of the layer, (see, e.g., Oliner, (1978)). Therefore the Love wave behaviour is determined by the layer (i.e. by the wave guide properties), for which the KdV equation appearance in long waves limit is quite natural.

On the contrary, the layer provides only dispersion effect for Rayleigh surface waves, and dispersion in such elastic system (thin layer upon half-space) is described by non-local Hilbert transform term, that is also may be anticipated, taking into account similar wave propagation problems for stratified liquids, considered by Benjamin and Ono.

We have obtained the final equations (5.96), (5.103) and (5.130), starting from the basic nonlinear elasticity theory. One should mention the phenomenological method (see, e.g., Korpel and Banerjee, (1984)) applied to the nonlinear strain waves study in layered half-space. However, the results obtained here allow us to point out some restrictions of that method. First of all, the coefficients of the equations in (Ewen et al., 1982) cannot be defined analytically. Moreover the final form of longitudinal strains equation does not depend on the type of boundary conditions at the layer–half-space interface. The procedure of nonlinear term coefficient phenomenological determination, proposed in (Ewen et al. 1982; Korpel and Banerjee, 1984), could

---

[2]The scanned images of old oscillogrammes did not provide reasonable quality in figures.

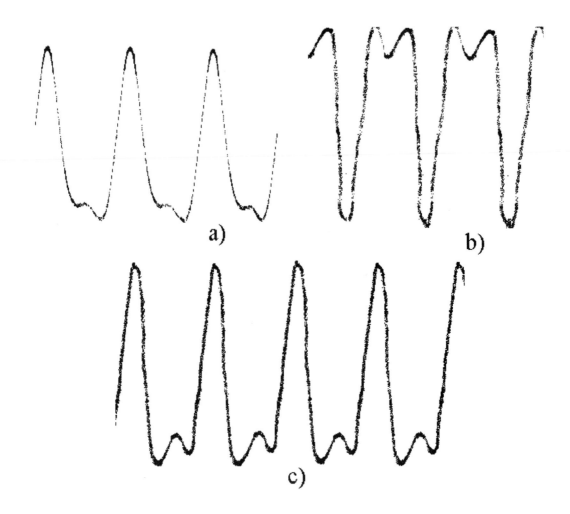

Figure 5.17: Oscillograms of the shape of the Rayleigh wave moving from the right in the following intervals: *a*) 9.04-9.5 mm; *b*) 9.24-9.75 mm from a transducer, and: *c*)at the distance 9.46 mm from the transducer. After Nayanov (1986).

be applied to some extent, when the envelope solutions are considered, because both contact models are governed by the same NLS equation. However, when the base-band solutions are studied, the differences in nonlinear terms in (5.96) and (5.130) are of considerable importance, and there is no solution to (5.130), which is similar to those obtained for the Benjamin-Ono equation (5.96).

One of the main conclusion to be made is in the fact that, depending upon the type of the interface contact model, the governing equations possess qualitatively different travelling wave solutions for nonlinear elastic strains.

5.3.6 On physical experiments in waves in a layered medium

The periodical wave shapes observed in various experiments reported in (Dyakonov et al. 1988) are surprisingly similar to those calculated above in the framework of the slippage contact model. It allows the suggestion that in an experiment the *full interface contact*, achieved initially due to the metal layer evaporation onto a substrate during the manufacturing, being subjected to the powerful acoustic pulse loading, is transformed thereafter to the *sliding contact*.

Physical applications of this idea to the superconductivity experiments in the thin lead film coated the $LiNbO_3$ substrate (see Dyakonov et al. 1988a,b) can be used to explain both threshold area variation and control. It was found there, in particular, that the localized zone of elastic compression in lead results, indeed, in significant growth of the superconductive threshold zone's width in the lead layer. Comparing the periodical wave shapes in Figures 5.11-5.16, one can see that such localized zones can be observed only in Figure 5.11. The waves shown in the last figures describe the compression and extension strains of the same magnitudes, and therefore the influence of compression zone is compensated by the nearest extension zone influence. Moreover the wave shape, defined experimentally in (Dyakonov et al. 1988b) just before the layered area, is similar to the B-O periodical wave shape (5.97) (see Figure 5.11).

Therefore, according to the theory developed here, such zones may arise only when the layer slippage has been taken into account. The dramatic variation of elastic properties (i.e., in boundary conditions of the problem) of thin nonlinearly elastic layer could clarify also a film demolition phenomenon from a substrate, observed in experiments, see (Dyakonov et al. 1988a). The layer thickness-wave length ratio and the strains in this work were of the same order, then these data satisfy the choice of the small parameter ϵ introduced above. Only the qualitative estimations for experiments mentioned were done above, because the Murnaghan moduli values for both lead and lithium niobate were beyound reach.

The different distances between the interdigital transducer and the layered area given in (Dyakonov et al. 1988a; 1988b) and in (Cho and Miyagawa, 1993; Nayanov, 1986) may be the main reason of different shapes of the strain waves observed. The layered area was placed rather far from the transducer in (Dyakonov et al. 1988b), and the initial powerful harmonic surface acoustic wave has enough time to change its shape running to the layered area. As a result the wave shape became similar to the B-O periodical wave shape just before the layer, and may play the role of a suitable initial condition for generating the B-O wave (5.97) in the layer. On the contrary, the layered area is placed almost directly near the transducer in (Cho and Miyagawa, 1993; Nayanov, 1986), therefore a practically undeformed harmonic wave goes to the layer. That is why the different waves, describing by (5.105) or (5.104), arise, in our opinion, in these experiments.

Eventually, it seems to be of interest that the Murnaghan moduli values of the layer influence only on the nonlinear strain wave, when the *sliding contact* is assumed.

Vice versa, only the Murnaghan moduli values for the half-space define the strain wave evolution in the *full contact* problem. At the same time the linear Lamé constants of both the layer and the half-space are included in wave parameters relationships independently of the contact model considered. Therefore the so-called physical nonlinearity (characterized by any set of the 3d order elastic moduli, e.g., by the Murnaghan moduli) and geometrical nonlinearity (characterized by the finite elasticity tensor **C**) play essentially different roles in the elastic wave propagation in a layered system, depending on the chosen model of a contact.

To sum up, dealing with the long nonlinear wave propagation in a layered elastic half-space, we have studied two models of nonlinearly elastic layer–half-space contact.

We have shown that the Benjamin-Ono nonlinear equation obtained by means of the rigorous asymptotic analysis provides a description of longitudinal base-band nonlinear strain waves, when the contact between the layer and the half-space admits layer's sliding. The KdV approach claimed recently in several papers, can hardly be applicable to this nonlinear elastic wave problem. The envelope waves propagation may be governed by the NLS equation. Additional conditions were found, under which the long nonlinear periodical and solitary strain waves may exist in various materials.

For the full contact statement of the problem considered here we have derived the new complicated integro–differential equation and, pursuing an asymptotic solution of it, we reduced the problem to the NLS equation, having an envelope solitary strain wave solution, but not a conventional long nonlinear wave. The elastic compression area, e.g., responsible for the superconductivity threshold control, does not occur, if there is no slippage on contact surface.

The envelope wave solutions allows us to describe waves having the shapes similar to those observed in recent experiments and to propose their theoretical description.

The following description of physics of a wave propagation in layered half-space seems to be feasible. Being evaporated onto a substrate, the film is in full contact with a substrate at the beginning. Powerful acoustic loading initiated by transducers leads to the nonlinear wave propagation, governed by the equation obtained here, that has a solution in terms of an envelope soliton, in particular. The increase of the acoustic power leads to the variation of contact type, the slippage occurs, providing some local compression areas, and this stage is governed by the Benjamin-Ono equation for nonlinear strains. At this step the superconductivity thershold can be under acoustic (or, more precisely, elastic strains) control. However, further enhancement of stresses results in a nonstationary process of contact's erosion and demolition of a layer.

Chapter 6

Numerical simulation of the solitary waves in solids

This chapter is devoted to descriptions of various numerical experiments in propagation and the amplification of the strain solitary wave (soliton) in different nonlinearly elastic wave guides.

Despite adequate theoretical analysis of generation and propagation of solitary strain waves in rods and plates, even recently, serious difficulties have been met in multiple attempts to generate, detect and observe such waves.

We have shown that the strain soliton propagates without change of shape in uniform rod, layer or plate, while its shape will vary in presence of inhomogeneities. In the last case, the amplification or focusing may occur, in other words, the soliton amplitude will increase, while its width will decrease simultaneously. Then the localized area of plasticity and even fracture of a wave guide, may happen to appear. That can be important for applications and cannot be governed by any linear theory.

In particular, that was one of the reasons for thorough study of both theoretical statements and possible formal solutions of the problem, together with the estimations of influence of inhomogeneity, dissipation and types of elastic nonlinearity. Necessary refinements in the physical model of solitary strain wave propagation in solids were discussed above; now discussions will be aimed at considering some problems of numerical simulation of non-stationary and complex problems of strain solitary wave in solids which may hardly be studied analytically.

As we mentioned in the Introduction, to arrange a real physical experiment one has to estimate a lot of parameters of an incident wave pulse, an influence of inhomogeneity, impurity, dissipation, etc.

For this reason, based on the theory developed and experiments performed, the results of numerical simulation as the cheapest experiment in waves in complex wave guides will be well based and informative. Possible artifacts can be avoided using analytical study.

The complete description of a three-dimensional nonlinear wave in continuum is difficult, which is why initial 3-D problems are usually reduced to the 1-D form in order to clarify the simplest but qualitatively new *analytical* solutions. Very often the linearization of a problem was done, mostly in engineering problems, however, it turns out to be unsatisfactory from the genuine physical point of view, because the ratio of a finite deformation and its linear part is determined by a displacement gradient and its variation in time, see, e.g., Lurie, (1980); Pleus and Sayir, (1983); Shield, (1983); Engelbrecht, (1983), etc.

Successful experimental generation of a strain soliton in a rod with permanent or varying cross section discussed in Chapters 4 and 5 revived the interest to numerical simulation of strain solitons in complex solids, which may not be studied in another way up to this date.

In this Chapter we shall deal with numerical simulation of the propagation of a long nonlinear strain wave in different uniform or inhomogeneous wave guides.

6.1 Numerical simulation of non-stationary deformation waves

The numerical simulation of nonlinear waves in 1+1D problem (in rod) will be based mainly on the dimensionless form of the DDE, which we shall write now as

$$4(u_{tt} - u_{xx}) = \left[6u^2 + au_{tt} - bu_{xx}\right]_{xx}.$$

where coefficients were scaled in order to avoid a small parameter in simulation. Its influence can be recovered easily by scaling, while in computations we do not need an asymptotics. It was shown that this equation has an exact solution in the form of a solitary wave

$$u(x, t) = \alpha \cosh^{-2} \sqrt{\frac{\alpha}{a - b + a\alpha}} \left(x \pm \sqrt{1 + \alpha}\right) \tag{6.1}$$

which will be the basic one for numerical experiments. The variation of characteristics of the wave guide results in the DDE with variable coefficients, therefore we consider also a generalised version of the equation that may be written as

$$4u_{tt} - g(x)u_{xx} = a(x)u_{xxtt} + b(x)u_{xxxx} + e(x)u_{xxx} + d(x)u_{xxt} + F(u, u_x, u_{xx}, u_{xxx}) \tag{6.2}$$

where all functions $a(x), b(x), d(x), e(x), g(x)$ continuously depend on x, and residual term $F(\cdot)$ does not contain time derivatives. The discretisation of (6.2) yields

$$
\begin{aligned}
4u_{\bar{t}t}(x_i, t_{n+1}) - g(x_i) u_{x\bar{x}}(x_i, t_{n+1}) = {} & a(x_i)u_{\bar{t}t x\bar{x}}(x_i, t_{n+1}) + b(x_i)u_{x\bar{x}x\bar{x}}(x_i, t_{n+1}) + \\
& e(x_i)u_{x\bar{x}(x)}(x_i, t_{n+1}) + d(x_i)u_{\bar{t}x\bar{x}}(x_i, t_{n+1}) + \\
& F(u, u_{(x)}, u_{x\bar{x}}, u_{x\bar{x}(x)}; x_i, t_{n+1})
\end{aligned} \tag{6.3}
$$

where subscripts x, and $\bar{x}$, and (x) denote the forward, backward, and separated finite differences respectively. All approximations of derivatives are calculated in one grid point (x_i, t_{n+1}) with no use of the derivatives at the previous steps, and the approximation accuracy is the same for all terms, therefore the discrete version (6.3) is self-consistent.

The standard approach to the solution of the discrete version of quasi hyperbolic nonlinear equation governing the nonlinear wave propagation problem is based, as a rule, on a solution to the 3-diagonal matrix linear equation

$$AU^{n+1} = F^n \qquad (6.4)$$

written for a mesh function $U^{n+1} = \{u_j\}^{n+1} = \{u(x_j, t_{n+1})\}$ with coefficients

$$A_{j,j} = 2 + 4h^2/a; \; A_{j,j-1} = A_{j,j+1} = -1, \; j = 0 \dots M \qquad (6.5)$$

and the right hand side term

$$F_j^n = h^2 \left(u_{j,x\bar{x}}^{n-1} - 2u_{j,x\bar{x}}^n \right) - \frac{4h^2}{a} \left(u_j^{n-1} - 2u_j^n \right) + \frac{h^2\tau^2}{a} \left[c\left((u_j^n)^2 \right)_{x\bar{x}} - b\left(u_j^n \right)_{x\bar{x}} \right)_{x\bar{x}} + 4u_{j,x\bar{x}}^n \right] \qquad (6.6)$$

where subscripts x and $\bar{x}$ denote forward and backward finite differences, respectively.

The numerical problem (6.4-6.6) is solved as a rule using the Godunov implicit scheme (Godunov, Ryabenky, 1973). Stability of the approach defines by a relation

$$Q_j = \frac{A_{j,j}}{A_{j,j-1} + A_{j,j+1}} = 1 + \frac{2h^2}{a} \begin{cases} > 1, a > 0 \\ < 1, a < 0 \end{cases}$$

In the simplest model of strain soliton propagation in the uniform rod a parameter $a \propto \nu^2 > 0$ and therefore the implicit finite difference scheme is stable.

For a model governed by the modified DDE, having different coefficients, one has $a \propto \nu(\nu - 1) < 0$, $b \propto \nu c_0^2 > 0$, and $|a| < |b|$. In this case $Q_j < 1$ and the scheme described above diverges, in spite of the fact that we have obtained for the model a set of various stable analytical solutions, also in form of solitons[1].

Therefore it was necessary to modify an approach in order to make a numerical scheme stable for *any* set of coefficients. To begin with, we shall rewrite (6.3) as

$$-\frac{2a(x) + \tau d(x)}{2h^2\tau^2} u_{i+1,n+1} + \left[\frac{4}{\tau^2} + \frac{2a(x) + \tau d(x)}{2h^2\tau^2} \right] u_{i,n+1} - \frac{2a(x) + \tau d(x)}{2h^2\tau^2} u_{i-1,n+1} =$$

$$= \xi \left[u\left(x_i, t_{n+1} \right) \right] + \eta \left[u\left(x_i, t_n \right), u\left(x_i, t_{n-1} \right) \right],$$

where the new operator $\xi(\cdot) = g(x)D_{xx} + b(x)D_{xxxx} + e(x)D_{xx(x)} + F(\cdot)$ contains no time derivatives, D is the central difference operator, and

$$\eta[\cdot] = -\frac{1}{h^2\tau^2} \left[2u_{x\bar{x}}\left(x_i, t_n \right) - u_{x\bar{x}}\left(x_i, t_{n-1} \right) \right] - \frac{1}{2h^2\tau} u_{x\bar{x}}\left(x_i, t_{n-1} \right) -$$
$$\frac{4}{\tau^2} \left[-2u\left(x_i, t_n \right) + u\left(x_i, t_{n-1} \right) \right]$$

contains the discrete time derivatives at the steps n and $n - 1$.

---

[1]It is typical for nonlinear dynamics problems and underlines clearly a necessity of analytical solutions as the 'check points' for numerical simulation.

Now we consider a converging iterative scheme for (6.3) with arbitrary values of coefficients. The approach is based on an assumption that a solution of (6.2) is a solution of an equation for $u(x,t;\xi)$:

$$\rho u_\xi + 4u_{tt} - g(x)u_{xx} = a(x)u_{xxtt} + b(x)u_{xxxx} + e(x)u_{xxx} + d(x)u_{xxt} + F(u, u_x, u_{xx}, u_{xxx}),$$
(6.7)

stationary with respect to ξ. Here ρ is an artificial dissipation coefficient. Rewiting the equation as

$$\rho u_\xi = f(u),$$

one can easily demonstrate the main idea. For simplest $f(u) = u + C(x,t)$ we have

$$u(x,t;\xi) = C(x,t) + \exp\left[\frac{D(x,t)}{\rho}\xi\right]$$

where $D(x,t)$ is independent of ξ. Therefore both the collapsing solution for growing ξ ($\mathrm{sgn}D(x,t)\rho = 1$) and the stationary one ($u^\infty(x,t,\xi) = C(x,t)$) may be obtained using the appropriate $\mathrm{sgn}\rho$.

The discrete version of (6.7) for successful iterations has the form:

$$Bu_{n+1}^{(k+1)} = \xi\left[u_{n+1}^{(k)}\right] + \eta\left[u_n^{(\infty)}, u_{n-1}^{(\infty)}\right] + \frac{\rho}{\varsigma}u_{n+1}^{(k)}$$
(6.8)

where the components of the 3-diagonal matrix B are

$$B_{p,p+1} = B_{p+1,p} = -\frac{2a(x_p) + \tau d(x_p)}{2h^2\tau^2}; \quad B_{p,p} = \frac{4}{\tau^2} + \frac{2a(x_p) + \tau d(x_p)}{2h^2\tau^2} + \frac{\rho}{\varsigma}$$
(6.9)

and ς is a mesh size for ξ, while a superscript (∞) denotes the limiting value of iterative sequence. The condition of convergence for the 3-diagonal matrix , according to (Il'yin, 1995) has the form:

$$\frac{(4 + \rho\tau^2/\varsigma)h^2}{2a(x_p) + \tau d(x_p)} > 0, \forall p,$$

that allows to define ρ and ς:

$$\begin{cases} \rho = 0, & 2a(x_p) + \tau d(x_p) > 0 \\ \rho < -4\varsigma/\tau^2, & 2a(x_p) + \tau d(x_p) \le 0 \end{cases}$$

and to conclude that the grid size for ξ is to be $\varsigma = O(\tau^2)$. Eventually, we define $r = \rho/\varsigma$ and obtain a condition for a proper choice of the normalised dissipation coefficient r:

$$\begin{cases} r = 0, & 2a(x_p) + \tau d(x_p) > 0 \\ r = -4\varsigma/\tau^2 - v, & 2a(x_p) + \tau d(x_p) < 0 \end{cases}$$
(6.10)

where any small v lies in the interval $0 < v \ll 1$. Values of r below the limits in (6.10) result in deterioration of convergence, therefore the scheme proposed is always convergent if a solition exists for given $r, a(x_p), \tau, d(x_p)$. Starting values of u for each step will be

$$u_{n+1}^{(0)} = u_n^{(\infty)},$$

because $\tau \ll 1$. At each step of ξ the system of linear equations with one and the same matrix B will be solved by means of the LU-decomposition adopted for the 3-diagonal system, as it was proposed in (Godunov, Ryabenky, 1973).

Verification of the simulation routine is based on the conservation laws resulted from physics and comparison of numeric solution to (6.2) with the soliton (6.1). Optimal choice of mesh sizes in x and t is based on the consequent decrease of steps in them and on the conservation laws. In particular, the mass conservation is enough for estimation of h, while for τ the energy conservation law is used. For example, for the soliton propagation problem in a uniform rod the values $h = \lambda/15$, $\tau = \lambda/(20V_0)$ are good enough (error was less than 1%) for the $O(h^2, \tau)-$ scheme, where λ is a width of the pulse calculated in its half-height, and the soliton velocity is defined as $V_0 = \sqrt{1 + \alpha} = \sqrt{(\lambda^2 - b)/(\lambda^2 - a)}$.

The preliminary results of numerical simulation have shown that the mesh of $O(h^2, \tau)$ is quite satisfactory for most of calculations, however the scheme with $O(h^4, \tau)$ is more effective and reliable for estimation of characteristics of a solitary wave propagating along a wave guide with rapidly varying parameters and/or the varying coefficients of the linear wave operator which requires very small grid size in space.

For numerical simulation of wave processes governed by the equation (6.2) we use implicit finite difference schemes of orders $O(h^2, \tau)$ and $O(h^4, \tau)$ where h is a mesh size in space and τ - in time, and the grid is uniform. The DDE, in contrast to the KdV equation, is quasi hyperbolic and describes a wave motion in two opposite directions, therefore central differences are used for approximation of derivatives. The schematic of the grid $O(h^2, \tau)$ is shown in Figure 6.1.

The contunuous boundary conditions for localized solution (6.1) will be changed in numerical simulation in

$$u(\pm L/2) = 0; \quad u_x(\pm L/2) = 0$$

where L is the length of a wave guide along the wave propagation direction.

To develop the stable scheme for an equation with $a < 0$, $b < 0$ and $|a| < |b|$ we substitute the second derivative in space by means of linear wave operator $u_{xx} = u_{tt} + O(\epsilon)$. After substitution of it into the DDE, containing ϵ, and negletion of terms of $O(\epsilon^2)$ we obtain the Boussinesq equation of the kind:

$$4u_{tt} - 4u_{xx} = \left[6u^2 + |b - a|u_{tt}\right]_{xx},$$

for which the implicit scheme (6.4-6.6) with new coefficients $a \rightarrow |b - a|$, $b \rightarrow 0$ is always stable. Again ϵ was excluded for simplicity. The difference between an analytical solution and numerical solution found via this modified scheme will be comparable with the accuracy of the initial problem statement.

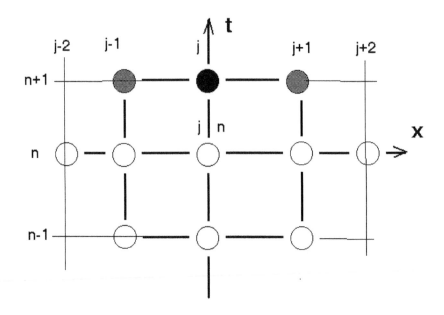

Figure 6.1: Grid of central differences of $O(h^2, \tau)$

6.1.1   Parametrisation of the strain solitary wave

To decrease the dimension of the discrimination problem let us consider the parametrisation of a strain solitary wave based on the characteristics that were observed in experiments, see e.g., (Dreiden et al. 1988). Following (Bukhanovsky and Samsonov, 1998), we introduce several quantities defined with respect to the wave profile function $u(x, t)$:

1) Amplitude of the main pulse as a global maximum of the wave profile:

$$A_0(t) = \max_x \left[|u(x, t)|\right]. \tag{6.11}$$

2) Phase shift of the main pulse equal to the difference between the maxima of both observed and standard (defined for an uniform wave guide) wave profiles:

$$\Delta\varphi = \left\{ x_0 \colon u(x_0, t) = A_0(t) \right\} - \left\{ x_e \colon u_e(x_e, t) = A_e(t) \right\}. \tag{6.12}$$

3) Characteristic length of the main pulse defined as the distance $\lambda = x_2 - x_1$, where:

$$x_1 = u^{-1}(A_0/2), \quad x_1 < x_0, \qquad\qquad x_2 = u^{-1}(A_0/2), \quad x_2 \geq x_0. \tag{6.13}$$

4) Parameter of '*peakedness*' of the main pulse, positive for the steeping of the front slope and negative for the steeping of back one:

$$\vartheta(t) = \int_{x_1}^{x_0} u(x, t)dx - \int_{x_0}^{x_2} u(x, t)dx. \tag{6.14}$$

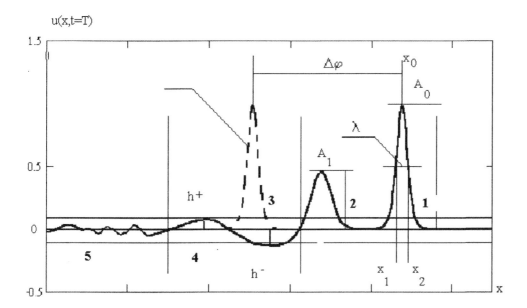

Figure 6.2: Parametrisation of a solitary wave and shelf behind it. Numbers in the field denote : 1) and 2) - two consequent solitary waves; 3)- soliton in a uniform rod; 4) - a shelf ; 5) - an oscillating wave package.

Also the auxiliary qualitative characteristics of the wave may be considered:

1) The positive or negative shelf behind the main pulse.

2) The train of secondary strain solitons behind the main pulse.

3) The oscillating wave package behind the main pulse..

4) The reflected oscillating wave package that propagated in inverse direction from the main pulse.

The parameters of the strain solitary wave are shown schematically in Figure 6.2.

Note that in the figures here and below the thickness of the rod is in no correlation with the width or any other pulse parameters, otherwise a rod image would be just a line. The pulse width, in fact, may be equal to dozens of radii of a rod.

The values of the above described parameters depend strongly on the (hyper)elastic properties of a wave guide and its geometry. It is convenient to define the qualitative classification for the strain solitary waves in a rod, having a circular cross-section, based on the behaviour of $R(x)$ as follows:

· Uniform wave guides, having constant properties;

· Geometrically homogeneous wave guides, $R(x) - const$;

· Wave guides, which cross section area $R(x)$ is monotonously increasing (decreasing);

· Wave guides with the periodic and quasiperiodic cross section area;

· Wave guides, having varying cross section area;

· Complex wave guides with $R(x)$ being a stepwise function;

· Wave guides, containing insertions with varying cross section area.

In addition, in a real wave guide, the profile of a solitary wave may be different from any of the above mentioned due to roughness and erosion of the wave guide surface. The influence of these phenomena are considered separately.

6.2   Solitary waves in a homogeneous rod

Before we start with numerical solutions of complicated soliton propagation problems, we shall verify the numerical simulation approach by the simplest problem of strain soliton propagation in a homogeneous wave guide. For the investigation of the non-stationary process of strain soliton generation the numerical scheme defined in (6.3), (6.4), (6.8), (6.9) is used.

To begin with, the generation of two strain solitons due to the symmetric decay of stepwise sharp initial pulse was simulated, as it is shown in Figure 6.3.

The stepwise pulse decay is accompanied by the high frequency oscillating wave package due to the high gradient values at the boundaries of initial pulse. Due to dispersion, the amplitude of oscillating package decreases and the total length increases, so the mass conservation integral is constant. The stabilization of the amplitude of the main pulse will not provide a criterion of stationarity because the steady value of the amplitude appears faster than of the length (see Figure 6.3-II and III). The distinctive criterion of stationarity (i.e., the generation of strain soliton) is the soliton separation from the oscillating package (see Figure 6.3-III). The linear wave's velocity (a " tail" behind the soliton) is less than of the main pulse, and the interaction between the tail and the main pulse becomes negligible after the soliton separation. Formally it corresponds to the stabilization of the energy conservation integral value.

As well as the KdV equation, the DDE has the multisoliton solutions. The generation of three-soliton train from a wide initial stepwise pulse is shown in Figure 6.4.   The strain solitons generation depends on elasticity of the wave guides, and primarily, on the balance between dispersion and nonlinearity in order to provide the stable solitary waves. Without the balance, the initial pulse is transforming into the package of linear waves due to dispersion, as shown in Figure 6.5 in the problem of a decay of initial stepwise pulse with the $A_0 < 0$ (extension wave) in the wave guide with the $sgn(b) = -1$ (compression solitary wave allowed).

There is no stable wave structure, and the amplitude of the main pulse decreases monotonously, while the main part of the nonstationary wave has the same velocity as the oscillating package behind it. Hence, due to continuous interaction between the main pulse and "a tail" in Figure 6.5-III the amplitude of the main pulse becomes almost equal to the amplitude of oscillations behind it.

One of the distinctive features of soliton solutions is their interaction stability. In Figures 6.6-6.7, interactions of two different solitons are shown: the head-on collision and the passing collision. After the interactions, both shape and amplitude of the solitons remain practically unchanged, however the phase shift and an oscillating wave package occur behind the solitons, because the DDE is not fully integrable by the inverse scattering transform method.

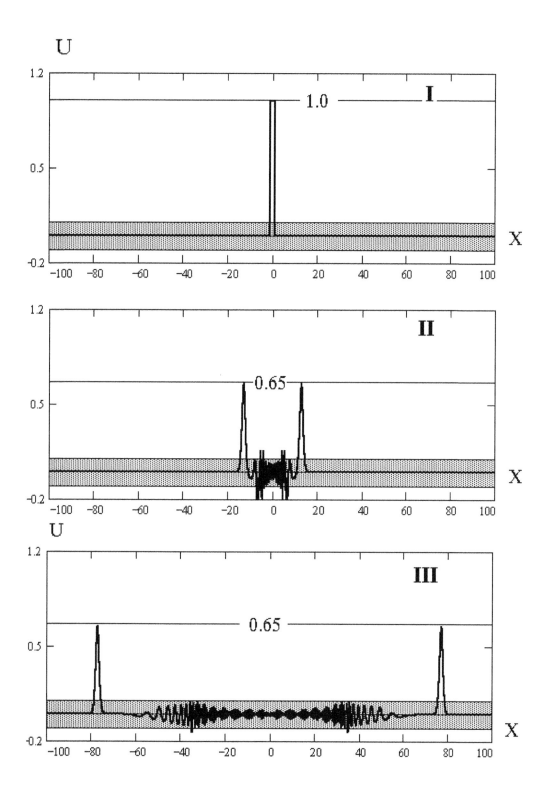

Figure 6.3: The solitary wave formation from a sharp pulse.

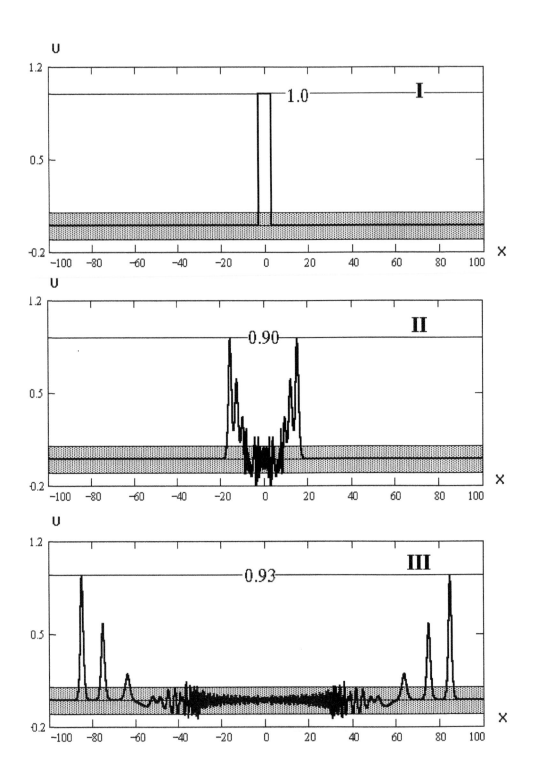

Figure 6.4: The formation of the train of solitons from an initial massive pulse.

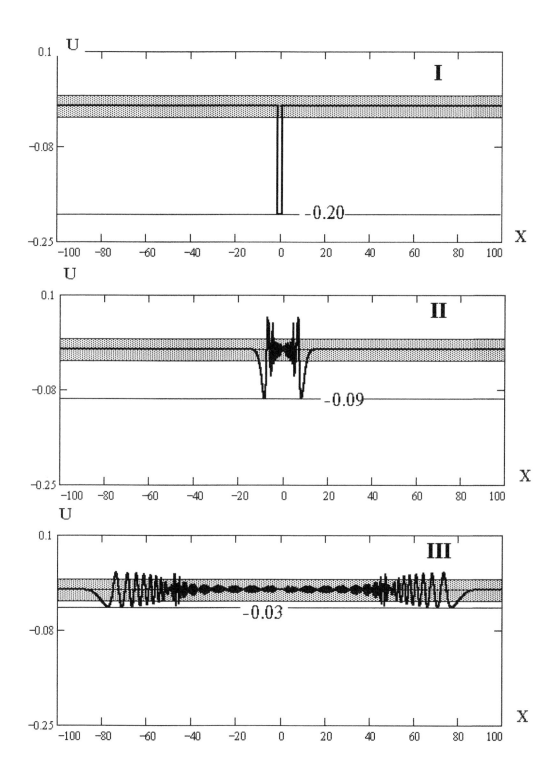

Figure 6.5: Decay of an extension wave in a rod having an inappropriate sign of nonlinearity for soliton propagation.

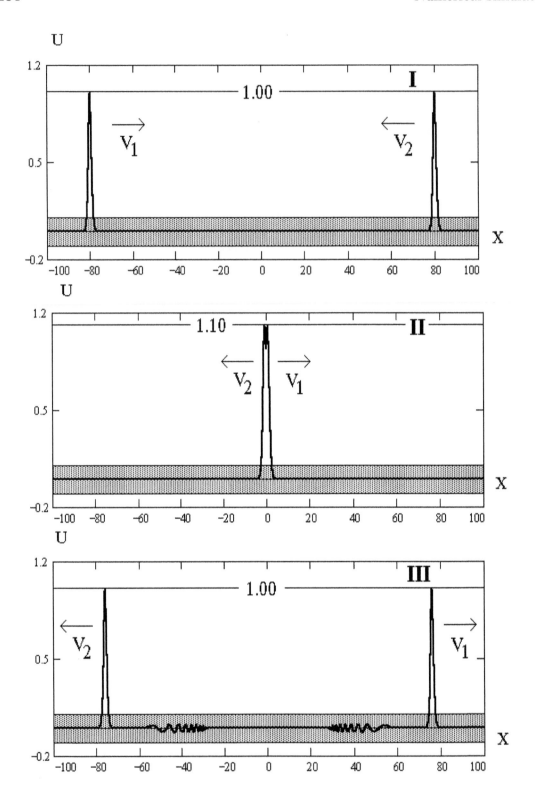

Figure 6.6: Head-on collision of the DDE solitons.

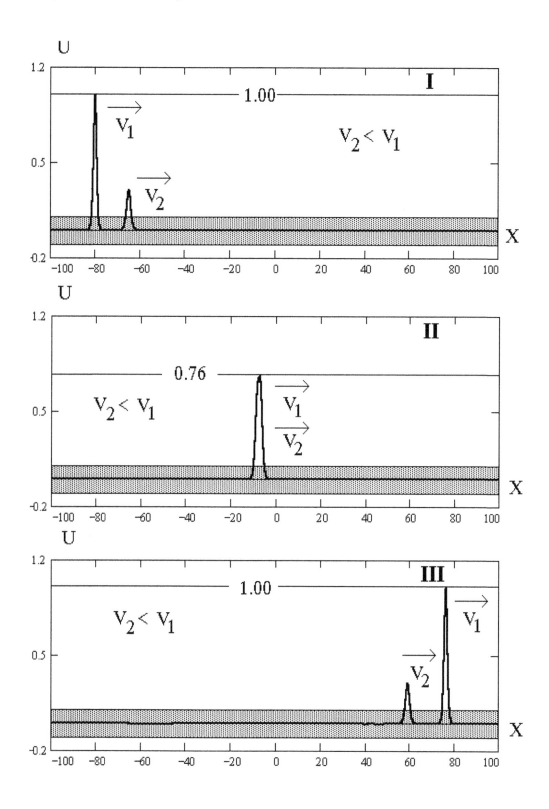

Figure 6.7: The passing collision of the DDE solitons.

The energy conservation integral remains practically constant before and after the solitons collision, however during the interaction its stationarity fails (it increases for the head-on collision and decreases for passing collision). Hence, as for the KdV equation (Lamb, 1983), the amplitude of the main pulse in the head-on collision increases, while it decreases for the passing collision.

For many applications, the interaction of the strain soliton and a linear wave may be of interest as a demonstration of stability of the strain solitary wave in the interaction with noise signals, e.g., with reflected linear waves in real physical experiments. We have modelled it as the interaction between the soliton and the part of the steady harmonic pulse $W(x) = \alpha \cos [\omega x]$ with no initial velocity and positive mass, as it is shown in Figure 6.8. After the collision the soliton remains practically unchanged, but only a weak oscillating package is generated. In the same time the initial steady harmonic pulse was transformed to the moving train of new small solitons (because the total mass of the initial pulse is positive) by means of the nonlinear wave "filter" governed by the DDE.

The parameters of the numerical experiments performed with solitary waves propagating in homogeneous wave guide are shown in Table 6.1.

Table 6.1. Parameters of the numerical simulation of the strain soliton propagation in the uniform rod.

Figure no.	a	b	L	α_1	α_2	$m(t)$	$S_1(t)$	l
6.3	0.5	0.2	200	1.0	–	6.0	3.35	6.0
6.4	0.5	0.2	200	1.0	–	2.0	1.24	2.0
6.5	0.5	0.2	200	–0.2	–	–0.4	–	2.0
6.6	0.5	0.2	200	1.0	1.0	3.58	2.39	1.41
6.7	0.5	0.2	200	1.0	0.3	2.52	1.34	1.41
6.8	0.5	0.2	200	1.0	0.3	1.19	–	1.41

Here: l is the length of initial pulse, α_i are the amplitudes of the first (and second, if any) initial pulses, $m(t)$ is the value of the mass integral, $S_1(t)$ is the steady value of the energy integral, a and b are the values of coefficients in the DDE.

6.3 Solitary waves in a nonuniform rod

Let us consider the main features of the strain solitary wave propagation in different geometrically inhomogeneous wave guides. To begin with, we shall deal with the solitary waves in wave guides having smoothly varied cross section area.

In Figure 6.9 the soliton propagation is shown in a tapered wave guide, which radius $R(x)$ decreases linearly, while the polynomial smoothing in the vicinity of a switching point between uniform and narrowing sections is applied. The condition $S''(x) \in C[L]$ is valid, where L is the total length of the wave guide. The parameters of the wave guide and the initial pulse are shown in Table 6.2 below.

One can see the steepness of the front of the soliton propagating in the tapered rod and smoothness of its back together with an increase of both amplitude and

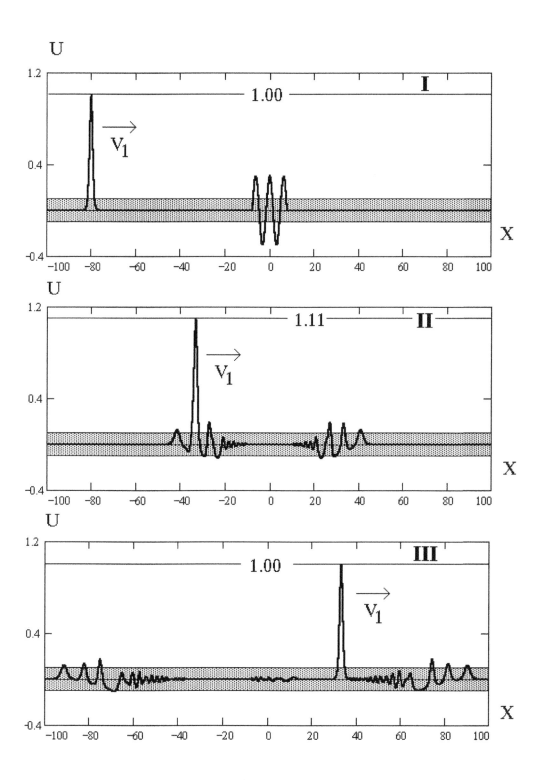

Figure 6.8: The soliton collision with a harmonic wave.

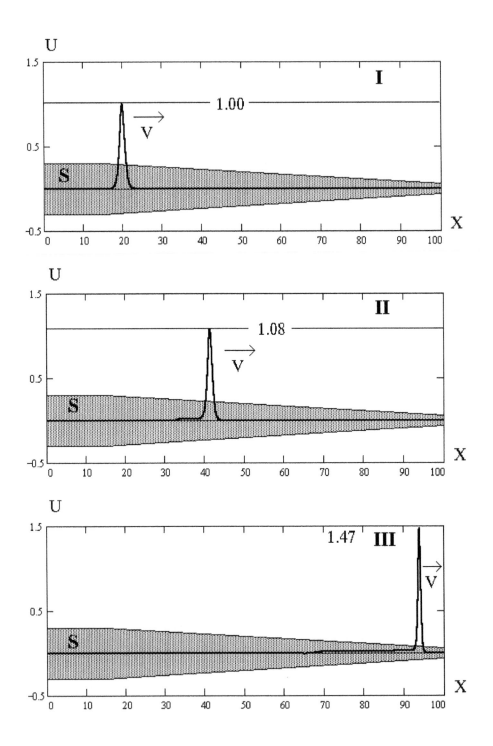

Figure 6.9: A soliton focusing in a tapered rod.

velocity of the soliton (see Figure 6.9.III), and a decrease of length of the main pulse, and the positive shelf generation behind the soliton. The velocity of the shelf wave is positive much less than the velocity of the soliton. These effects observed in the numerical simulation were confirmed by means of the experiment.

Vice versa, the back of the soliton propagating in a widening rod (Figure 6.10) is steep in comparison to its front, the velocity and amplitude decrease and the length increases. The negative shelf behind the soliton is generating and due to negative mass defect is transforming eventually into an oscillating wave package due to dispersion influence.

Let us consider now the problem of soliton sensitivity to the cyclic variation of $R(x)$ in order to confirm a resistence of the pulse with respect to it and a recurrence of the soliton parameters in the homogeneous rod after propagation in the inhomogeneous one. Geometry of the wave guide and results of numerical simulation are shown in Figures 6.11-6.12.

It is shown that the double (negative-positive) shelf is generating during the soliton propagation in the bottleneck-like (narrowing-widening) wave guide. The positive long strain wave behind a soliton generates due to propagation in a tapered section, the negative one - in the widening section. The variation of the soliton amplitude in the uniform section (after this cyclic change of a cross section, see Figure 6.11-III) is negligible in comparison to an amplitude of the initial pulse.

The behaviour of strain wave in widening and narrowing wave guide is quite similar, see Figure 6.12. The amplitude of the soliton decreases negligibly, while the shape of the shelf is opposite with that observed in Figure 6.11. The result of this study is: propagation of a soliton in a rod with smooth, finite and long variations of $R(x)$from and to a given value does not change its amplitude, but only a phase shift occurs together with small shelf behind of the main pulse. It means that such a wave guide is transparent for nonlinear localised strain wave. The energy of these elastic deformations will propagate along the wave guide without considerable losses.

6.4   Solitary waves in complex rods

Now we will consider the simulation of solitary waves propagation in complex wave guides containing some uniform sections with the different cross sections. Obviously, the behaviour of the main pulse may be similar to the results obtained for an inhomogeneous wave guide with the $R''(x) \in C[L]$, however the soliton transition through a jump in cross section generates the distinctive transient process.

In Figure 6.13 the soliton propagates in a complex rod from the wide section to the narrow one. Focusing of the soliton may be easily seen, and a small part of the initial pulse energy is reflecting from the jump in $R(x)$ between sections. The main pulse is 20% taller in the narrow section, and the train of two solitons occurs.

This is the main difference in comparison with the soliton propagation in a wave guide with the smoothly varied cross section: instead of a positive shelf wave a new soliton occurs.

During the soliton transition from the narrow section to the wide one, Figure 6.14

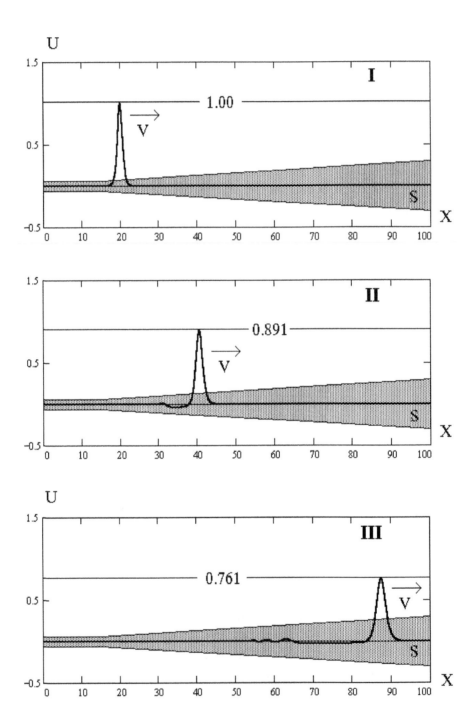

Figure 6.10: Attenuation of a soliton in a widening rod.

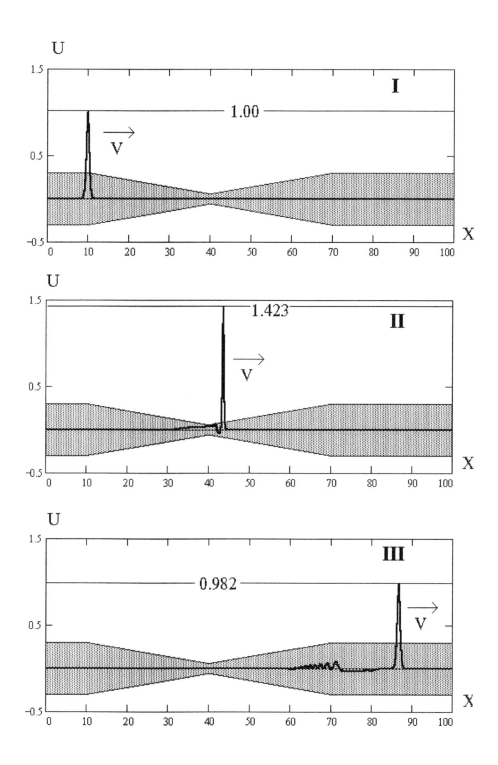

Figure 6.11: A soliton recurrence in a bottle neck rod.

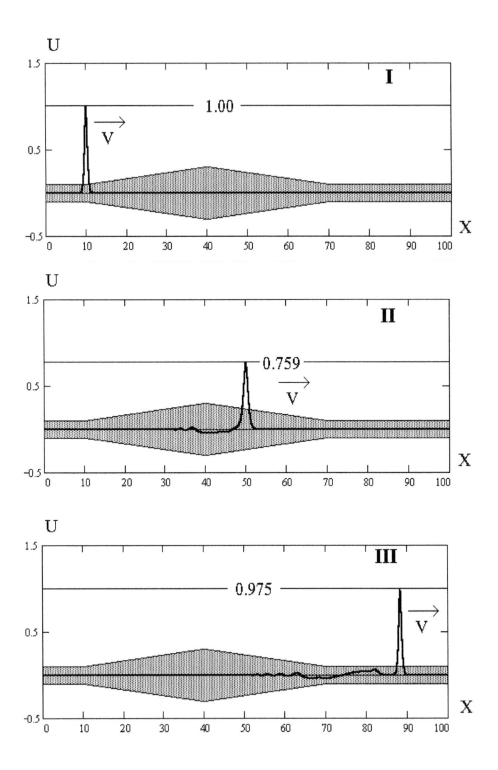

Figure 6.12: A soliton recurrence in a puffed rod.

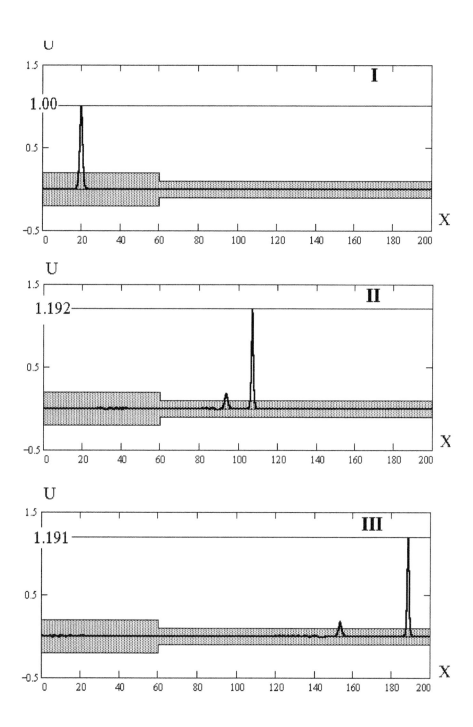

Figure 6.13: A soliton evolution due to transition from thick to thin part of rod.

its amplitude decreases almost the same 20%, while the second soliton does not occur - the negative shelf wave decays in the oscillating package due to the dispersion.

The next step is in consideration of the soliton propagation in a complex wave guide with narrow (Figure 6.15) or wide (Figure 6.16) insertion. The simulation results show the recurrence of both the soliton's shape and amplitude in the uniform section of the rod after an insertion, while the phase shift and small plateau behind the soliton are the only and definite signs of insertion presence. In Table 6.2, the main parameters of soliton simulation in complex rods are shown, where S_i is either the relative cross section length of the i-th uniform section or the length of the end cross section of smoothly varied rod, L_i is the length of the $i-$th section in parts of the total length L, and type of a cross section variation is indicated also. Parameters a and b were described above.

Table 6.2. Parameters of the numerical simulation of strain solitons in smoothly changing and complex rods.

Figure	a	b	L	Type	S_0	S_1	S_2	S_3	L_0	L_1	L_2	L_3
6.9	0.5	0.2	100	smooth	1.0	0.2	–	–	0.15	1.00	–	–
6.10	0.5	0.2	100	smooth	1.0	3.0	–	–	0.15	1.00	–	–
6.11	0.5	0.2	200	smooth	1.0	0.2	1.0	1.0	0.10	0.40	0.70	1.00
6.12	0.5	0.2	200	smooth	1.0	3.0	1.0	1.0	0.10	0.40	0.70	1.00
6.13	0.5	0.2	200	step	1.0	0.5	–	–	0.30	1.00	–	–
6.14	0.5	0.2	200	step	1.0	2.0	–	–	0.30	1.00	–	–
6.15	0.5	0.2	200	step	1.0	0.7	1.0	–	0.30	0.55	1.00	–
6.16	0.5	0.2	200	step	1.0	1.5	1.0	–	0.30	0.55	1.00	–

6.4.1 Strain solitons in a rod with the periodically varied cross section

In previous numerical experiments the strain soliton exhibited very strong resistance against any attempts to effect to its shape and amplitude, using the wave guide geometry variations. Only phase shift and shelf generation seems to be the irreversible signs of pulse penetration inside a rod with complex geometry. In fact, it was quite predictable, and will be useful for applications, e.g., in strikers.

Now we consider the soliton propagation in wave guides having periodically varied cross section area (PVCSA), e.g., in screws and similar structures.

Obviously, both processes of focusing and small shelf generation will be periodic in space, too. Taking into account the irreversibility of the shelf generation it might be expected that the soliton in the long wave guide with PVCSA would be attenuated and transformed eventually into a train of linear wave packages.

To study it let us approximate the centred cross section radius by means of equivalent harmonic model (Husu et al., 1975):

$$R(x) = A_R \cos \left[\frac{2\pi}{T_R} x + \varphi_R \right]. \tag{6.15}$$

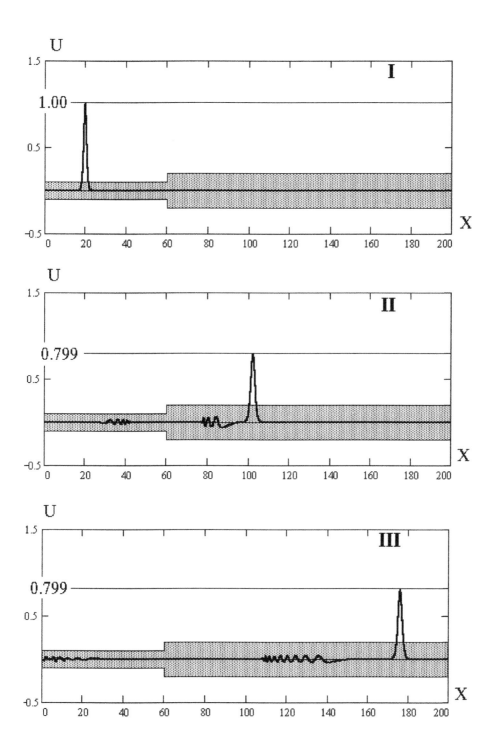

Figure 6.14: A soliton attenuation due to transition from the thin to the thick part of a rod.

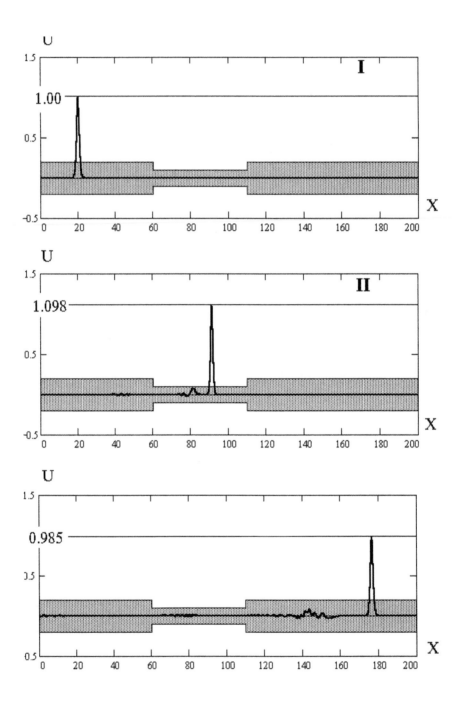

Figure 6.15: A soliton in a rod with thinner insertion.

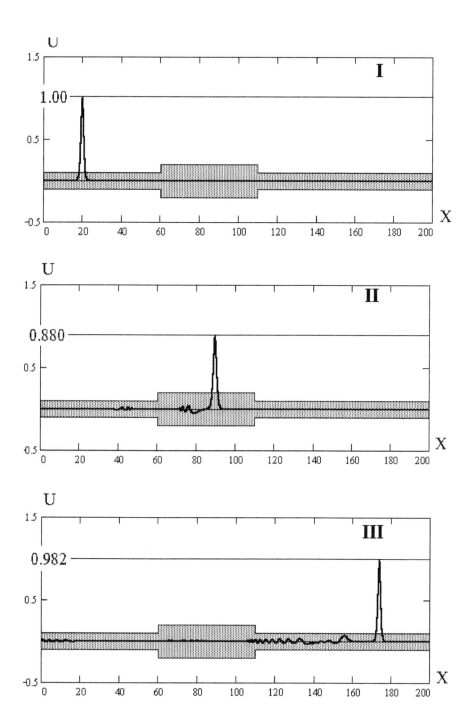

Figure 6.16: A soliton in a rod with thicker insertion.

In Figures 6.17-6.19 the soliton propagation inside insertions with PVCSA having different periods ($T_R = 5; 10; 18.75$ λ) are shown. Here $A_R = 0.1R_0$, where R_0 is the average radius of the wave guide.

For each case the soliton propagation through one cycle of the wave guide generates the positive (for narrowing part) and the negative (for widening part) shelves consequently that is similar to an oscillating " tail" . The number of oscillations is equal to the number of cross section area cycles.

The decrease of the amplitude of soliton due to irreversible losses of the energy for the shelves generation, as well as of the velocity of the pulse, is recognisable, however surprisingly small. The negative phase shift Δ is observed.

It is shown that the less the frequency $\omega_R = 2\pi/T_R$ of the cross section variation, the more the decrease of the amplitude and velocity, i.e., the soliton is less sensitive with respect to the high frequency oscillations of the cross section area. The soliton is more sensitive to the oscillations of the cross section area, having a period comparable with a pulse length measured at the half-amplitude level.

The amplitude A_R in (6.15) effects the shape of the solitary wave, too. In Figure 6.18 the solitary wave propagation in a wave guide with $A_R = 0.8R_0$ and $T_R = 10l$ (two times greater than for Figure 6.17) is shown.

It is seen that the phase and amplitude defects are greater, hence, the relative decay of the soliton is faster.

The amplitude of the main pulse $A_0(t)$ in the wave guide with PVCSA is harmonically decreasing due to irreversible energy losses in the main pulse and transfer to the shelves. In Figure 6.20 the space trajectory of the soliton amplitude variation $A_0(t) = u(x_0, t)$ for Figure 6.18, is shown; on can see practically linear decrease of the trajectory.

The parameters of the numerical simulation are shown in Table 6.3.

Table 6.3. Parameters of the numerical simulation of strain soliton propagation in the wave guides having the PVCSA.

Figure	a	b	L	T_R	A_R	ϕ_R	λ	A_0	L_R/T_R
6.17	0.5	0.2	200	$5l$	0.12	0	0.894	1.0	30
6.18	0.5	0.2	200	$10l$	0.12	0	0.894	1.0	15
6.19	0.5	0.2	200	$18.75l$	0.12	0	0.894	1.0	8

Here L is the length of the wave guide, L_R/T_R is the length of PVCSA insertion (as ratio from the T_R), A_0 is the amplitude of initial pulse, all other values were defined in (6.15).

6.4.2 An example of classification using soliton transformation

The simple qualitative classification may be done based on the results of numerical simulation of strain soliton generation and propagation in geometrically inhomogeneous wave guides.

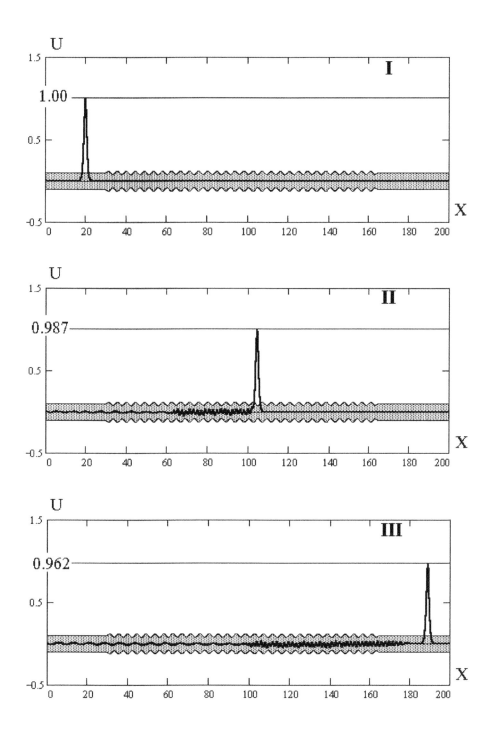

Figure 6.17: A soliton propagation in a rod with periodically varying cross section, $T_R = 5$.

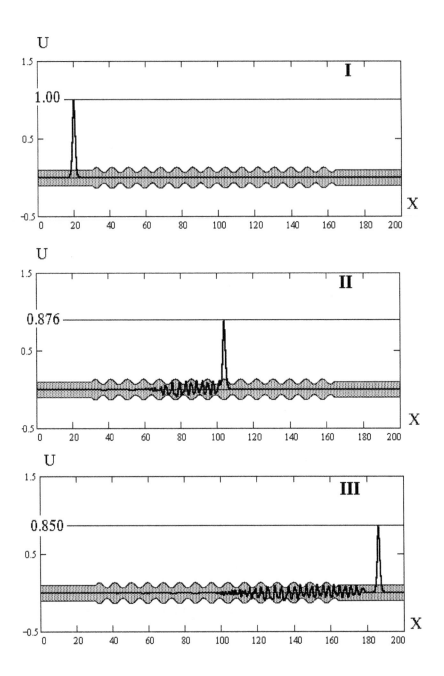

Figure 6.18: A soliton propagation in the same rod for $T_R = 10$.

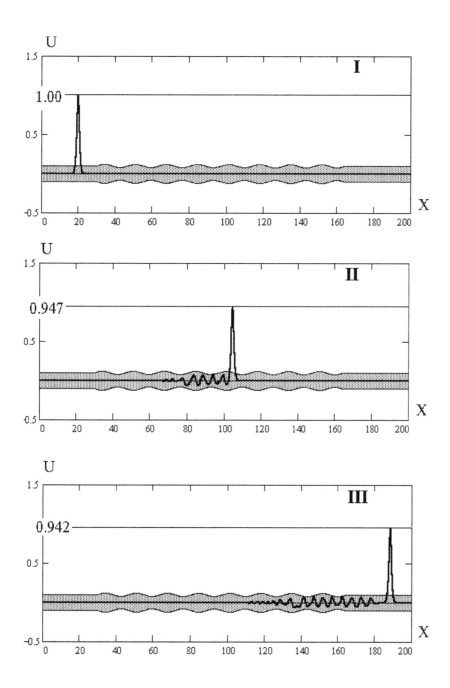

Figure 6.19: A soliton propagation in an insertion with periodically varying cross section area with large period $T_R = 18.75$.

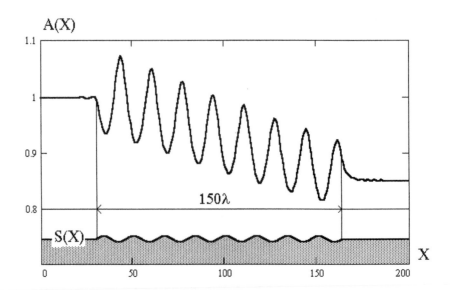

Figure 6.20: Space trajectory of the soliton amplitude variation corresponding to the soliton propagation in rods with a periodically varied cross section.

The classification allows us to propose the preliminary idea why a wave guide under consideration belongs among one of the above mentioned classes that were considered as typical for structural elements.

The characteristics of the strain solitary waves for the different classes of wave guides are shown in Table 6.4.

Table 6.4. Strain solitary waves parameters for the different classes of wave guides.

Rod	A_0/A_e	$\Delta\phi$	λ/λ_e	θ	presence of shelf	2nd soliton	reflection wave
uniform	1	0	1	0	0	0	0
narrowing	> 1	$+$	< 1	$+$	$+$	0	0
widening	< 1	$-$	> 1	$-$	$-$	0	0
narrowing insertion	> 1	$+$	< 1	0	0	0	0
widening insertion	< 1	$-$	> 1	0	0	0	0
narrowing-widening	< 1	$-$	< 1	0	$+/-$	0	0
widening-narrowing	< 1	$-$	> 1	0	$+/-$	0	0
thick to thin; Fig. 6.13	> 1	$+$	< 1	0	0	$+$	$+$
thin to thick; Fig. 6.14	< 1	$-$	> 1	0	$-$	0	$+$
thin insertion	< 1	$-$	< 1	0	$-$	$+$	$+$
thick insertion	< 1	$-$	> 1	0	$-$	$+$	$+$
$R\cos(wx)$, short rod	?	$+/-$	1	$+/-$	$N+/-$	0	0
$R\cos(wx)$, long rod	< 1	$-$	> 1	0	$N+/-$	0	0

Here the notation "$N + /-$" is used to note the train of N positive and negative shelves, 0 means no effect, $+$ means existence of a feature or positive value of a parameter.

We can see from Table 6.4 that the set of criteria is enough to provide a hypothesis about the type of the wave guide for wave guides with the simple geometry only. For the verification of more complicated hypothesis (e.g., of tentative PVCSA property) it is necessary to compare the quantitative values of the above mentioned parameters. These values may be obtained for any particular elastic material by means of the numerical simulation. Moreover, in practice the additional noise effects may take place due to erosion and roughness of a wave guide surface. For this case the wave guide may not be uniquely identified. Moreover, it is important to investigate the stability of the strain soliton with respect to tentative random variations of the cross section area.

Following Husu et al. (1975) we shall describe the random roughness of a lateral surface of the rod by means of the following sum of the function (6.15) and the Gaussian random process $\eta(x)$:

$$R(x) = A_R \cos\left[\frac{2\pi}{T_R}x + \varphi_R\right] + \eta(x). \tag{6.16}$$

To solve the soliton propagation problem in such rod the direct Monte Carlo simulation method was used which allowed us to obtain the distribution of both amplitude A_R and phase φ_R at each time step. Assuming a simple autoregressive model for simulation of Gaussian process $\eta(x)$ is valid

$$\eta(x) = \sum_{k=1}^{N} \Theta_k \eta(x_{i-k}) + \sigma_\chi^2 \chi_i,$$

where $\chi_i \in N(0,1)$ is the Gaussian white noise, the parameters Θ_k of autoregression and the white noise variation σ_χ^2 are to be obtained from the Yule-Walker equations:

$$K(X_i) = \sum_{l=1}^{N} \Theta_1 K(X_{i-l}), \ \ X_i = x_p - x_r, \ \ p - r = i; \ i \in 0, ... N,$$

where $K(X_{i-l})$ is the covariant function of stationary process $\eta(x)$. The prescribed condition $R''(x) \in C[L]$ is equivalent to the inequality $N > 2$.

The result of numerical simulation of the soliton propagation in the rod with randomly disturbed (e.g., due to corrosion) lateral surface seemed to be almost predictable: the strain soliton decay is very small even for a correlation period equal to the pulse length λ and the recurrence of the soliton takes place when it goes along a uniform rod after the damaged part. Being a bulk density wave, the strain soliton is not very sensitive with respect to random disturbances of the lateral surface of the wave guide even in a long time propagation. However, the phase shift distribution is more sensitive to random disturbances. To sum up, in contrast to the linear wave propagation in a rod with randomly damaged lateral surface, the strain soliton does not have a remarkable decay even for disturbances with a correlation period similar to the pulse width. Passing on to the uniform rod, the soliton is recovering and propagating with velocity greater than a velocity of the plateau behind it. The phase shift may be informative for non-destructive testing of the rod, which lateral surface is randomly damaged.

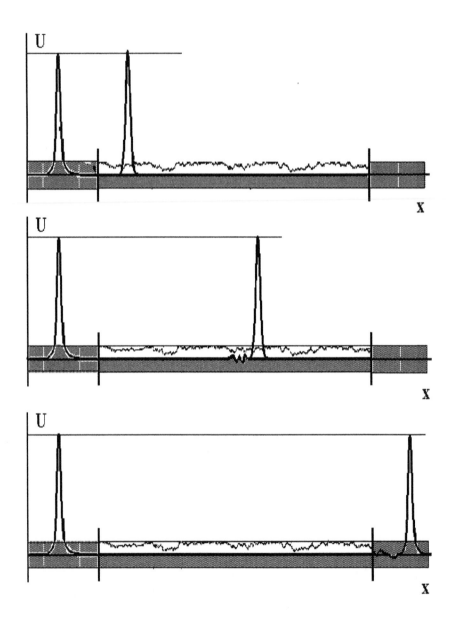

Figure 6.21: Propagation in a randomly corroded rod; soliton remains almost unchanged.

Conclusive remarks and tentative applications

Results of our theoretical, numerical and experimental attempts to investigate the strain solitons in solids allow us to conclude that we observed the main features of solitary waves, dealing with the long elastic strain waves in solids. They are:

1. **There is no long wave of opposite sign behind a soliton.**

   ⋆ Proved in theory, numerical simulations and experiments with real materials.

2. **A solitary wave shape remains permanent when propagating.**

   ⋆ Proved in theory, numerical simulations and experiments with real materials. It was observed at distances 20-30 times greater than a characteristic size in the problem, namely, the wave guide radius. Note that linear and even shock waves disappeared on these distances.

3. **A solitary wave is focused in a tapered wave guide.**

   ⋆ Proved in theory, numerical simulations and experiments with real materials. Amplitude and pulse width are proportional to the wave velocity in contrast to linear wave parameters. Phase shift keeps the data on wave guide parameters variation. The experimental arrangement also allowed us to prove the steepening of wave front and smoothening of its back.

4. **A soliton keeps the permanent shape even after a head-on collision with another soliton.**

   ⋆ Proved in numerical simulations. No interaction was observed with a linear or shock wave in a head-on collision. A small oscillating tail was recognised in simulations behind a solitary wave after collision. The study of reflection allowed the preparation of the experiments in head-on collisions.

We hope for some interesting applications of the nonlinear wave phenomena that were discussed here. Besides many references already mentioned throughout the book, we shall mention the influence of nonlinearity of dynamic processes in elasticity discovered in behaviour of various structures. One was mentioned by Coste and Montes (1986) and Montes (1987). They dealt with optical transmission line and studied the stimulated Brillouin backscattering (SBS) generated by the coupling of the electromagnetic wave $\lambda = 1.06 \mu m$ with the thermal acoustic fluctuations of thin optical fiber.

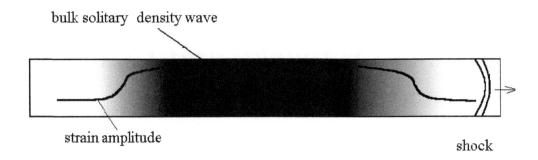

Figure 6.22: The long solitary density wave in a rod.

They observed that a backward elastic wave profile begins with a *peak of very large amplitude* followed by several secondary peaks, that the dynamics of the 1st peak is *insensitive* to dissipation and reported the following mechanical effects on the fused silica fiber: strong electric field $E_{pulse} = 10^9 V/m$ provides the pressure $p = 200\ atm$ close to the fiber fracture threshold $e_0 \approx 350\ atm$. They confirmed that the linear elasticity theory is not satisfactory, and the elastic compression wave limits the laser pulse energy used for transmission due to nonlinear elasticity of deformation of thin glassy and plastic fibers. Another example is reported by Harker (1988) in application to non-destructive testing of prolonged pipelines, where it was shown the losses in signal transmission allowing choise of bulk compression waves, however, such linear waves will suffer losses of intensity through mode conversion to shear waves at each reflection. The solitary and other nonlinear bulk waves in solids attracts the growing interest in applications, see e.g., Babitsky ed., (1998), Engelbrecht (1991), Gurbatov et al. (1991), Jeffrey and Engelbrecht eds., (1994), Kivshar and Malomed (1989), Maugin (1988), Mayer (1995), Nesterenko (1992) and many others.

In particular, solitary waves of density may be useful in:

• Nondestructive testing and evaluation to construct macrostructure tests in aerospace and automotive structures, pipelines, machinery, where ultrasonic measurements could need weeks of work, as well as in various problems of introscopy or the microstructure tests of ceramics, optical fibres, crystalline, amorphous and glassy substances, alloys, dislocations and defects.

Vice versa, the solitary strain waves of permanent shape and power are good candidates for:

• Deformation energy transfer that may be used to improve various anti-shock structures: small shock can provide a soliton generation and, consequently, a hazardous energy transfer without loss in a working area; or

• Use in shock-transparent materials applicable in chemical reactors and other places where a reversible energy pulse is necessary to start another process.

• Evaluation of high order elastic moduli in material sciences because materials behaviour is highly nonlinear, essentially under dynamic loading. There is a remarkable discrepancy between measured, evaluated and predicted values of material constants, while precise measurements are required to avoid typical errors, which are up to 250% now.

• Impact manufacturing in engineering and applied physics for the contactless penetration, for design of strikers for hard, resistive and diamond-like materials, plastics and glass, because the repeated loading is possible without plastic flow of material of striker.

However, a number of new problems are to be solved in each tentative area of application. We have to note that the whole story about solitons in solids was based on proper theoretical concepts. To begin with new problems, it is necessary to develop the methods to solve nonlinear p.d.e. and corresponding boundary value problems, to design the 3D simulation and 3D visualization packages of various nonlinear waves in elastic and viscoelastic solids, to improve the computation methods for elastic moduli for various crystals, to find solutions to dissipative and non-local nonlinear wave propagation problems in different wave guides, in crystalline, porous, granular and amorphous solids.

Bibliography

[1] Ablowitz, M. and Segur, H. (1981). *Solitons and Inverse Spectral Transform.* SIAM, Philadelphia.

[2] Amiranoff, F. et al. (1985). Laser-driven shock wave studies using optical shadowgraphy. *Phys. Rev. A*, 32, 6, 3535-3546.

[3] Antman, S. (1998). *Nonlinear Elasticity.* Springer-Verlag, Berlin.

[4] Arrigo, D.J., Broadbridge, P. and Hill, J.M. (1993). Nonclassical symmetry solutions and the methods of Bluman-Cole and Clarkson-Kruskal. *J. Math. Phys.*, 34, 10, 4692-4703.

[5] Babitsky, V.I., ed. (1998). *Dynamics of Vibro-impact Systems.* Springer-Verlag, Berlin.

[6] Barenblatt, G.I. (1982). *Intermediate asymptotics.* Gosmeteoizdat, Moscow, *in Russian.*

[7] Bataille, K. and Lund F. (1982). Nonlinear waves in elastic media. *Physica D.* 6, 1, 95-104.

[8] Bateman, G. and Erdelyi, A. (1953-54). *Higher Transcendental Functions*, vol.3, McGraw Hill, New York.

[9] Bell, J.F. (1973). Experimental foundation of solid mechanics. In: *Encyclopaedia of Physics*, YIa/1, S.Flügge ed., Springer-Verlag, Berlin.

[10] Benjamin, T.B., Bona, J.G., and Mahoney, J.J. (1972). Model equations for long waves in nonlinear dispersive systems. *Phil. Trans. Roy. Soc.*, 272A, 47-78.

[11] Benjamin, T.B. (1966). Internal waves of finite amplitude and permanent form. *J. Fluid. Mech.* 25, 241.

[12] Berkovich, L.M., Nechaevsky, M.L. (1983). On group features and integrability of the Emden-Fowler type equations. In: *Group theoretical methods in physics*, vol. 2, Nauka, Moscow, 463-471, *in Russian.*

[13] Bharatha, S. and Levinson M. (1977). On physically nonlinear elasticity. *J. of Elasticity*, 7, 3.

[14] Biryukov, S.V., Gulyaev, Yu.V., Krylov, V.V. and Plessky, V.P. (1991). *Surface Acoustic Waves in Inhomogeneous Media*. Nauka, Moscow, *in Russian*.

[15] Bishop, A.R. (1979). Solitons in condensed matter physics. *Physica Scripta*, 20, 409-423.

[16] Bland, D.R. (1969). *Nonlinear Dynamic Elasticity*. Blaisdell Publishing Co., Waltham, MA.

[17] Blatz, P.J. (1969). Application of large deformation theory to the thermo-mechanical behavior of rubberlike polymers-porous, unfilled and filled. In: *Rheology - Theory and Applications*, vol.5, Eirich, F.R., ed., Academic Press, pp.1-55.

[18] Bluman, G.W., Cole, J.D. (1974). *Similarity Methods for Differential Equations*, Springer-Verlag, Berlin.

[19] Bogardus, E.H. (1965). Third-order elastic constants of Ge, MgO and fused SiO_2. *J. Appl. Physics*, 36, 8, 2504-2513.

[20] Bogdanov, A.N. and Skvortsov, A.T. (1992). Nonlinear elastic waves in a granular medium. *J. Physique*, IY, 2, C1779-1782.

[21] Born, M. and Wolf, E. (1964). *Principles of Optics*. 2nd ed., Pergamon Press, Oxford.

[22] Boutroux, P. (1920-21). On multiform functions defined by differential equations of the first order, *Ann. of Math.*, 22, 1-10.

[23] Breazeale, M. and Philip, J. (1984). Determination of 3d order elastic constants from ultrasonic second harmonic generation measurements. In: *Physical acoustics XVII*, Mason, W.P. and Thurston R.N., eds., Academic Press, New York, 1-60.

[24] Bullough, R.K. and Caudrey, P.J., eds., (1980). *Solitons*. Springer-Verlag, Berlin.

[25] Bushman, A.V., et al. (1996). Experimental study of phenilone and polystyrene under shock loading. *JETP*, 109, 5, 1662-1670.

[26] Calogero, F. and Degasperis, A. (1982). *Spectral Transform and Solitons*, North-Holland, Amsterdam.

[27] Cantrell, J.H. (1989). Acoustic nonlinearity parameters and higher order elastic constants of crystals. *Proc IOA*, 11, Part 5, 445-452.

[28] Cho, Y., Wakita, J., and N. Miyagawa, (1993). *Japan J. Appl. Phys.*, 32, 2261.

[29] Cho, Y., and Miyagawa, N. (1993). Surface acoustic soliton propagating on the metalic grating waveguide. *Appl. Phys. Lett.*, 63, 1188.

[30] Chree, C. (1889). The equations of an isotropic elastic solid in polar and cylindrical coordinates, their solutions and applications. *Trans. Phil. Soc. Cambridge*, vol. 14, 250.

[31] Clarkson, P.A. and Kruskal, M.D. (1989), New similarity reductions of the Boussinesq equation. *J. Math. Phys.*, 30, 10, 2201-2213.

[32] Clarkson, P.A., LeVeque, R.J., and Saxton, R. (1986) Solitary wave interactions in elastic rods. *Studies in Appl. Math.*, 75, 2, 95-122

[33] Clarkson, P.A. and Winternitz, P. (1991), *Physica D*, 49, 257 .

[34] Cole, J.D. (1968). *Perturbation Methods in Applied Mathematics*. Blaisdell Publishing Co., Waltham, MA.

[35] Coste, J. and Montes, C. (1986). Asymptotic evolution of stimulated Brillouin backscaterring: implication for optical fibers. *Phys. Rev. A.*, vol. 34, 5, 3940.

[36] Crandall, S.H. (1974). Nonlinearities in structural dynamics. *Shock and Vibration Dig.*, 6, 8, 1-13.

[37] Crawford, F. (1970). *Waves*. Berkeley physics course. vol. 3, McGraw Hill, Montreal.

[38] Crighton, D.G. (1995). Applications of KdV. *Acta Applicandae Mathematicae*, 39, 39-67.

[39] Dodd, R. et al. (1984). *Solitons and Nonlinear Wave Equations*, Academic Press, London.

[40] Drazin, P.G. and Johnson, R.S. (1989). *Solitons: an Introduction*. The University Press, Cambridge.

[41] Dreiden, G.V. et al. (1986). Interaction of shocks with plane interface of liquid and solids. *Sov. Techn. Phys. Lett*, vol. 12, 19, 1153-1158.

[42] Dreiden, G.V. et al. (1988). Observation of boundary conical waves in a liquid near lateral surface of elastic rod. *Sov. Techn. Phys. Lett*, vol. 14, 4, 310-313.

[43] Dreiden, G.V. et al. (1988). Formation and propagation of strain solitons in non-linearly elastic solids. *Sov. Phys. Techn. Phys.* 33, 10, 1237-1241.

[44] Dreiden, G.V. et al. (1989). Shock waves near liquid-solid interface. *Sov. Phys. Techn. Phys.*, 34, 1, 203-208.

[45] Dreiden, G.V., Ostrovskaya, G.V., and Ostrovsky, Yu.I. (1994). Optical research into hydrodynamic liquid-phase processes initiated by laser radiation. *Thermophysics and aeromechanics*, 1, 2, 143-156.

[46] Dreiden, G.V. et al. (1995). Experiments in the propagation of longitudinal strain solitons in a nonlinearly elastic rod. *Tech. Phys. Letters*, 21, 6, 415-417.

[47] Dreiden, G.V. et al. (2001). Reflection of longitudinal strain soliton from the end of nonlinearly elastic rod. *Techn. Physics J.*, to appear.

[48] Dubrovsky, G.V. and Kozachek, V.V. (1995). Microscopic description of growth kinetics of thin films. *J. Techn. Physics*, 65, 4, 124-141, *in Russian.*

[49] Dyakonov, K.V., Ilisavsky, Yu.V., and Yakhkind, E.Z. (1988a). The influence of sound on superconductive lead films. *Sov. Techn. Phys. Lett.*, 14, 976.

[50] Dyakonov, K.V., Ilisavsky, Yu.V. and Yakhkind, E.Z. (1988b). Nonlinear effects into surface acoustic waves propagation in $LiNbO_3$ at $T = 300 - 4.2K$. *Sov. Techn. Phys. Lett.* 14, 12, 943.

[51] Editorial: Soliton wave receives crowd of admirers. *Nature*, vol. 376, 3 Aug 1995, 373.

[52] Eilbeck, J.C. (1978). Numerical studies of solitons. In: *Solitons and Condensed Matter Physics*, Bishop, A. and Schneider, A. eds., Springer-Verlag, Berlin.

[53] Elisseyev, V.V. (1988). On nonlinear dynamics of elastic rods. *J. Appl. Mathem. Mech.-PMM*, 52, 4, 635-641, *in Russian.*

[54] Engelbrecht, J. and Peipman, T. (1992). Nonlinear waves in a layer with energy influx. *Wave Motion*, vol. 16, 173-181.

[55] Engelbrecht, J. (1991). *Introduction to Asymmetric Solitary Waves.* Longman, London.

[56] Engelbrecht, J. and Nigul, U. (1981). *Nonlinear Deformation Waves.* Nauka, Moscow, *in Russian.* (see also: Engelbrecht, J. (1983). *Nonlinear Wave Processes of Deformation in Solids.* Pitman, London)

[57] Eringen, A.C. and Maugin, G.A. (1990). *Electrodynamics of Continua*, Springer-Verlag, New York.

[58] Eringen, A.C. and Suhubi, E.S. (1974). *Elastodynamics I, Finite Motions.* Academic Press, New York.

[59] Erofeyev, V.I. (1999). *Wave Processes in Solids with Microstructure.* The University Press, Moscow, *in Russian.*

[60] Erofeyev, V.I. and Potapov, A.I. (1993). Longitudinal strain waves in nonlinearly elastic media with coupled stresses. *Int. J. Non-lin. Mechanics*, 28, 4, 483-488.

[61] Ewen, J.F., Gunshor, R.L. and Weston, V.H. (1982). An analysis of solitons in surface acoustic wave devices. *J. Appl. Phys.*, vol. 53, 8, 5682 .

[62] Faddeev, L.D. and Takhtadzhan, L.A. (1986). *Hamiltonian Approach in Soliton Theory.* Nauka, Moscow, *in Russian.*

[63] Filonenko-Borodich, M.M. (1940). Some approximate theories of elastic foundation. *Uchenye zapiski MGU, Mekhanika*, vol. 46, 3-18, *in Russian*.

[64] Frantsevich, I.N., Voronov, F.F. and Bakuta, S.A. (1982). *Elasticity Constants and Moduli of Metals and Nonmetals*. Naukova Dumka, Kiev, *in Russian*.

[65] Frenkel, J. and Kontorova T. (1938). On the theory of plastic deformation and twinning. *Z. Physik Sowjet Union*, 13, 1.

[66] Fusco, D. and Oliveri, F. (1989). Derivation of a nonlinear model equation for wave propagation in bubbly liquids. *Meccanica*, 24, 1, 15-25.

[67] Fusco, D. (1994). Evolution equations and reduction approaches for nonlinear waves in solids. In: *Nonlinear Waves in Solids*, Jeffrey, Engelbrecht eds., Springer-Verlag, Wien.

[68] Golubev, V.V. (1941). *Lectures in the Analytical Theory of Differential Equations*. GITTL, Moscow-Leningrad, *in Russian*.

[69] Grigolyuk, E.S. and Selezov, I.G. (1973). *Non-classical Theories of Oscillations of Rods, Plates and Shells*. VINITI, Moscow, *in Russian*.

[70] Gulyaev, Yu.V. and Polsikova, N.I. (1978). *Sov. Phys. Acoustics*, 24, 287.

[71] Gundersen, R. M. (1990). Travelling wave solutions of nonlinear partial differential equations. *Int. J. Engng. Math.*, 24, 323-341.

[72] Gundersen R.M. (1992). Modified dispersive equations, I, II. *Int. J. Non-lin. Mechanics*, 27, 5, 651-660, 749-758.

[73] Gurbatov, S.N., Malakhov, A.N. and Saichev, A.I. (1990) *Nonlinear Random Waves in Nondispersive Media*. Nauka, Moscow, *in Russian*.(see also Gurbatov, S.N., Malakhov, A.N. and Saichev, A.I. (1991). *Nonlinear Random Waves and Turbulence in Nondispersive Media*. The University Press, Manchester).

[74] Gusev, B.E. and Karabutov, A.A. (1991). *Laser optoacoustics*. Nauka, Moscow, *in Russian*.

[75] Harith, M.A. et al. (1989). Dynamics of laser driven shock wave in water, *J. Appl. Phys.*, 66, 11, 5194-5197.

[76] Harker, A.H. (1988). *Elastic Waves in Solids with Applications to Nondestructive Testing of Pipelines*. Hilger & Bristol, Philadelphia.

[77] Hermite, C. (1873). *Course Lithographie de l'Ecole Polytechnique*, Paris, *in French*.

[78] Hunter, J.K. and Scheurle, J. (1988). Existence of perturbed solitary wave solutions to a model equation for water waves. *Physica D*, 32, 253-268.

[79] Ince, E. (1964). *Ordinary Differential Equations*. Dover, New York.

[80] Jeffrey, A. (1982). Acceleration wave propagation in hyperelastic rods of variable cross-section. *Wave Motion*, 4, 173-180.

[81] Jeffrey, A. and Engelbrecht, J., eds. (1994). *Nonlinear Waves in Solids*. Springer-Verlag, Wien.

[82] Jeffrey, A. and Kawahara, T. (1982). *Asymptotic Methods in Nonlinear Wave Theory*. Pitman, London.

[83] Jeffrey, A. and Xu, S. (1989). Exact solutions to the Korteweg-de Vries-Burgers' equation. *Wave Motion*, 11, 559-564.

[84] Johnson, R.S. (1970). A nonlinear equation incorporating damping and dispersion. *J. Fluid Mech.*, v.42, 49.

[85] Kadomtsev, B.B. and Petviashvili, V.I. (1970). On the stability of solitary waves in weakly dispersive media. *Soviet Physics -Doklady*, 15, 539

[86] Kalyanasundaram, N. (1987). Nonlinear mode coupling between Rayleigh and Love waves on an isotropic layered half-space. p. 47. In: *Recent Developments in Surface Acoustic Waves*, Parker, D.F. and Maugin G.A., eds., Springer-Verlag, Berlin.

[87] Kapaev, A. (1988). Asymptotics of solutions to the 1st order Painlevé equations. *Diff. Equations*, vol. 24, 10, 1684-1695, *in Russian*.

[88] Karabutov, A.A., et al. (1984). Direct observation of shock acoustic wave formation in solids. *Vestnik Mosk. Univ., ser.3, Fizika, Astronomiya*, vol. 25, 3, 89-91, *in Russian*.

[89] Karpman, V.I. and Maslov, E.M. (1982). Comments on "Soliton propagation in slowly varying media", *Phys. Fluids*, 25, 9, 1686-1687.

[90] Karpman, V.I. and Maslov, E.M. (1978). *JETP*, 75, 504.

[91] Kashcheev, V.N. (1990). *Heuristic Approach to Obtain Solutions to Nonlinear Equations of Solitonics*, Zinatne, Riga, *in Russian*.

[92] Kerr, A.D. (1964). Elastic and viscoelastic foundation models. *J. Appl. Mech.*, vol. E31, 491-498.

[93] Kivshar, Yu.S. and Malomed, B.A. (1989). Dynamics of solitons in nearly integrable systems. *Rev. Modern Phys.*, 61, 4, 763-915.

[94] Kodama, J. and Ablowitz, M.,(1981). Perturbation of solitons and solitary waves. *Stud. Appl. Math.*, vol. 64, 225-245.

[95] Ko, K. and Kuehl, H.H. (1978). Korteweg - de Vries soliton in a slowly varying medium, *Phys. Rev. Lett.*, 40, 4, 233-236.

[96] Konno, K. and Jeffrey, A. (1984). The loop soliton. In: *Advances in Nonlinear Waves*, Debnath, L. ed., Pitman, 162-182.

[97] Kolsky, H. (1963). *Stress Waves in Solids*. Dover, New York.

[98] Korpel, A. and Banerjee, P. (1984). Heuristic guide to nonlinear dispersive wave equations and soliton-type solutions. *IEEE*, 72, 1109.

[99] Korteweg, D.J. and de Vries, G. (1895). On the change of form of long waves advancing in a rectangular channel and on a new type of long stationary waves. *Phil. Mag.*, 5, 39, 422-443.

[100] Kudryashov, N.A. (1988). Exact soliton solutions to general evolution equation of wave dynamics. *Appl. Math. Mech.-PMM*, 52, 3, 465-470, *in Russian*.

[101] Kunin, I.A. (1975). *Theory of Elastic Media with Microstructure*. Nauka, Moscow, *in Russian*. Translation: (1982). Springer-Verlag, Berlin.

[102] Kurdyumov, S.P. et al. (1981). Dissipative structures in trigger media. *Differentsialnye Uravneniya*, 17, 10, 1875-1885, *in Russian*.

[103] Lakes, R.S. (1987). Foam structures with a negative Poisson's ratio. *Science*, 235, 1038-1040.

[104] Lakes, R.S. (1992). No contractile obligations. *Nature*, 358, 713-714.

[105] Landa, P. (1996). *Nonlinear Oscillations and Waves in Dynamical Systems*. Kluwer, Dordrecht.

[106] Landau, L.D. and Rumer, Yu.B. (1937). On sound absorbtion in solids. *Physikalische Z. Sowjetunion*, 11, 18.

[107] Lemke, H. (1920). *Sitzungsberichte Berlin Math. Gesell.*, 18, 26.

[108] Lie, S. (1883). Klassifikation und integration von gewoennlichen diffrentialgleichungen zwischen x,y, die eine gruppe von transformationen gestattet. III. *Archive von Math.*, Bd.YIII, H. 4, 371-458, *in German*.

[109] Lighthill, M.J. (1965). *J. Institute of Math. and Appl.*, 1, 1, 1-28.

[110] Lomov, S.A. (1981). *Introduction to the General Theory of Singular Perturbations*. Nauka, Moscow, *in Russian*.

[111] Love, A.E.H. (1952). *A Treatise on the Mathematical Theory of Elasticity*, 4th ed., Cambridge University Press, Cambridge.

[112] Lurie, A.I. (1980). *Nonlinear Theory of Elasticity*. Nauka, Moscow. *in Russian*. (English edition , 1990, Elsevier, Amsterdam).

[113] Lyamshev, L.M. and Sedov, L.V. (1981). Optical generation of sound in a liquid, *Acoust. Zhurnal*, 27, 1, 5-23, *in Russian*.

[114] McNiven, H.D. and McCoy, J.J. (1974). Vibrations and wave propagations in rods. In: *R.Mindlin and Applied Mechanics*, Herrmann, G., ed., Pergamon, NY, 197-226.

[115] Malmquist, I. (1914). Sur le fonctions à un nombre fini de branches définies par les équations différentielles du premier ordre. *Acta Math.*, 36, 310, *in French*.

[116] McIntosh, I. (1990). Single phase averaging and travelling wave solutions of the modified Burgers-Korteweg-de Vries equation. *Phys. Letters A*, 143, 1-2, 57-61.

[117] Maradudin, A.A. (1987). Nonlinear surface acoustic waves and their associated surface acoustic solitons. In: *Recent Developments in Surface Acoustic Waves*, Parker, D.F., and Maugin, G.A. eds., Springer-Verlag, Berlin, 62.

[118] Maugin, G.A. (1988). *Continuum Mechanics of Electromagnetic Solids.* Elsevier, Amsterdam.

[119] Maugin, G.A. (1994). Physical and mathematical models of nonlinear waves in solids. In: *Nonlinear Waves in Solids*, Jeffrey, Engelbrecht eds., Springer-Verlag, Wien.

[120] Maugin, G.A. and Hadouaj, H. (1990). Nonlinear surface transverse waves on elastic structures. In: *Frontiers of Nonlinear Acoustics: Proceedings of 12th ISNA*, Hamilton, M.F. and Blackstock, D.T., eds., Elsevier, London, 565.

[121] Mayer, A.(1995). Surface acoustic waves in nonlinear elastic media. *Physics Reports*, 256, 257-366.

[122] Mooney, M. (1940). A theory of large elastic deformation. *J. Appl. Phys.*, 11, 582-592.

[123] Molotkov, I.A. and Vakulenko, S.A. (1980). Nonlinear elastic waves in inhomogeneous rods. *Zapiski nauchn. semin. LOMI*, 99, 64-73, *in Russian*.

[124] Molotkov, I.A. and Vakulenko, S. A. (1988). *Concentrated Nonlinear Waves.* The University Press, Leningrad, *in Russian*.

[125] Montes, C. (1987). Inertial response to nonstationary stimulated Brillouin backscaterring: damage of optical and plasma fibers. *Phys. Rev. A.*, 36, 6, 2976.

[126] Morse, P.M. (1929). *Phys. Rev.*, vol. 34, 57.

[127] Murnaghan, F.D. (1951). *Finite Deformations of an Elastic Solid.* Wiley, New York.

[128] Murray, J.D. (1989). *Mathematical Biology.* Springer-Verlag, Heidelberg.

[129] Nariboli G.A. (1970). Nonlinear longitudinal dispersive waves in elastic rods. *J. Math. Phys. Sci.*, vol. 4, 64-73.

[130] Nariboli, G.A. and Sedov, A., (1970). Burgers's-KdV equation for viscoelastic rods and plates. *J. Math. Anal. Appl.*, vol. 32, 3, 661-677.

[131] Naugol'nykh, K.A. and Ostrovsky, L.A. (1990). *Nonlinear Wave Processes in Acoustics*. Nauka, Moscow, *in Russian*.

[132] Nayanov, V.I. (1986). Surface acoustic cnoidal waves and solitons in the structure of $LiNbO_3$ covered by SiO film. *JETP Lett.*, 44, 314-316.

[133] Nayfeh, A.H. (1973). *Perturbation Methods*. J.Wiley & Sons, New York.

[134] Nesterenko, V.N. (1992). *Pulse Loading of Heterogeneous Materials*, Nauka, Novosibirsk, *in Russian*.

[135] Newell, A. (1985). *Solitons in Mathematics and Physics*. SIAM, Philadelphia.

[136] Nicolis, G. and Prigogine, I. (1977). *Self-organisation in Non-equilibrium Systems*. J.Wiley & Sons, New York.

[137] Nikolova, E.G. (1977). *Soviet Physics JETP*, 45, 285.

[138] Nishinari, K. (1998). Discrete modelling of a string and analysis of a loop soliton. *J. Appl. Mech.*, 65, 9, 737-747.

[139] Nucci, M.C. and Clarkson, P.A. (1992). The nonclassical method is more general than the direct method for symmetry reductions. An example of the Fitzhugh-Nagumo equation. *Phys. Letters A*, 164, 49-56.

[140] Oliner, A.A. ed., (1978). *Surface Acoustic Waves*. Springer-Verlag, Berlin.

[141] Ono, H. (1975). Algebraic solitary waves in stratified fluids. *J. Phys. Soc. Japan*, 41, 1082.

[142] Ostrovsky, L.A. and Potapov, A.I. (1999). *Modulated Waves*. Johns Hopkins University Press, Washington.

[143] Ostrovsky, L.A. and Sutin, A.M. (1977). Nonlinear elastic waves in rods. *Priklad. Matem. i Mekhan.*, vol. 41, 3, 531-537, *in Russian*.

[144] Parker, D.F. (1984). On the derivation of nonlinear rod theories from three-dimensional elasticity. *J. Appl. Math. and Phys. (ZAMP)*, 35, 833-847.

[145] Parker, D.F. and Maugin, G., eds. (1987). *Recent Developments in Surface Acoustic Waves*. Springer-Verlag, Berlin.

[146] Parker, D.F., Mayer, A.P., and Maradudin, A.A. (1992). The projection method for nonlinear surface acoustic waves. *Wave Motion*, 16, 151-162.

[147] Parkes, E.J. (1994). Exact solutions to the 2D KdV-Burgers equation. *J. Physics A*, 27, L497-L501.

[148] Pasternak, N.L. (1954). *New Method for Calculation of Foundation on the Elastic Basement.* Gosstroiizdat, Moscow, *in Russian.*

[149] Pelinovsky, E.N., Engelbrecht, J. and Fridman, V.E. (1988). *Nonlinear Evolution Equations.* Longman, London,.

[150] Pleus, P. and Sayir, M. (1983). A second order theory for large deflections of slender beams. *J. Appl. Math and Phys. (ZAMP)*, vol. 34, 192-217.

[151] Pochhammer, L. (1876). Uber die Fortpflanzungsgeschwindigkeiten kleiner Schwingungen..., *J. f. Reine u. Angew. Mathematik (Crelle)*, vol. 81, 324, *in German.*

[152] Poincare, H. (1885). Sur un theoreme de M.Fuchs. *Acta Mathematica,* 1, 7, 6. *in French.*

[153] Porubov, A.V. and Samsonov, A.M. (1993). Refinement of longitudinal strain wave propagation in non-linearly elastic rod. *Sov. Tech. Phys. Lett.,* 19, 12, 365-366.

[154] Porubov, A.V. and Samsonov, A.M. (1994). Strain solitons on nonlinearly-elastic half-space coated by thin elastic film. In: *Abstract book, 2nd European Solid Mechanics Conf., Genoa,* S14 .

[155] Porubov, A.V. and Samsonov, A.M. (1995). Long nonlinear strain waves in layered elastic half-space. *Int. J. Non-linear Mechanics,* 30, 6, 861-877.

[156] Porubov, A.V. and Velarde, M.G. (2000). Dispersive-dissipative solitons in nonlinear solids. *Wave Motion,* 31, 197-207.

[157] Potapov, A.I. and Soldatov, I.N. (1984). Quasiplane nonlinear longitudinal wave beam in a plate. *Akusticheski zhurnal,* vol. 30, 6, 819-822, *in Russian.*

[158] Rabinovich, M.I. and Trubetskov, D.I. (1984). *Introduction in Oscillations and Wave Theory.* Nauka, Moscow, *in Russian.*

[159] Reissner, E. (1958). A note on deflections of plates on a viscoelastic foundation. *Trans. of ASME, J. Appl. Mech.,* vol.25, 144-145.

[160] Rosenau, P. (1986). Dynamics of nonlinear mass-spring chains near the continuum limit. *Physics Lett. A,* vol. 118, 5, 222-227.

[161] Rosenau P. (1988). Evolution and breaking of ion-acoustic waves. *Phys. Fluids,* vol. 31, 6, 1317-1319.

[162] Rudenko O.V. and Soluyan S.I. (1977). *Theoretical Foundations of Nonlinear Acoustics*, the Consultants Bureau, New York.

[163] Sakuma, T. and Nishiguchi, N. (1990). Theory of the surface acoustic soliton V. Approximate soliton solution of the Korteveg -de Vries type, *Phys. Rev. B,* 41, 12117.

[164] Samsonov A.M. (1982). Structural optimization in nonlinear elastic wave propagation problems. In: *Structural Optimization under Dynamical Loading. Seminar and Workshop for Junior Scientists*, U. Lepik ed., Tartu University Press, Tartu, Estonia, 75-76.

[165] Samsonov, A.M. (1984). Soliton evolution in a rod with variable cross section. *Sov. Physics-Doklady*, 29, 7, 586-587.

[166] Samsonov, A.M. (1987). Transonic and subsonic localized waves in nonlinearly elastic waveguides. In: *Proc. Intern. Conference on Plasma Physics*, Naukova Dumka, Kiev, vol. 4, 88-90.

[167] Samsonov, A.M. (1988a). Existence and amplification of solitary strain waves in non-linearly elastic waveguides. Leningrad, A. F. Ioffe Physical Technical Institute, Preprint no. 1259.

[168] Samsonov, A.M. (1988b). On existence of longitudinal strain solitons in an infinite nonlinearly elastic rod. *Sov. Phys.-Doklady*, vol. 33, 298-300.

[169] Samsonov, A.M. (1988c). On some exact solutions of nonlinear longitudinal wave equations with dispersion and dissipation. In: *Dispersive Waves in Dissipative Fluids*, Proc. EUROMECH Coll. 240, Bologna, D.G. Crighton and F. Mainardi eds., Tecnoprint, Bologna, 56-57.

[170] Samsonov, A.M. (1990a). Deformation waves in nonlinearly elastic inhomogeneous wave guides. *Atti Accad. Peloritana dei Pericolanti*, Messina, Italy, vol.68, s. 1, 535-551.

[171] Samsonov, A.M. (1990b). Nonlinear acoustic strain waves in elastic waveguides. In: *Frontiers on Nonlinear Acoustics*, M.F. Hamilton and D.E. Blackstock eds., Elsevier, Amsterdam, 588- 593.

[172] Samsonov, A.M. (1991) On some exact travelling wave solutions for nonlinear hyperbolic equations, In: *Nonlinear Waves and Dissipative Effects*, D. Fusco, A. Jeffrey eds., Longman, London, 123-132.

[173] Samsonov, A.M. (1993a) Some exact wave solutions in terms of the Weierstrass functions for nonlinear hyperbolic equations. In: *Future Directions in Nonlinear Dynamics*, P. Christiansen, C. Eilbeck, R. Parmentier eds., Plenum, New York, 125-128.

[174] Samsonov, A.M. (1993b) On exact travelling wave solutions for nonlinear acoustical problems with damping. In: *Advances in Nonlinear Acoustics*, H. Hobaek ed., World Scientific, London, 57-62.

[175] Samsonov, A.M. (1993c). Guided nonlinear elastic waves in films and plates: exact solutions. In: *IUTAM Symp. Nonlinear Waves in Solids, 15-20.08.93, Victoria, Canada. Abstracts of Invited Presentations*, 95.

[176] Samsonov, A.M. (1994) Nonlinear strain waves in elastic waveguides. In: *Nonlinear Waves in Solids*, A. Jeffrey and J. Engelbrecht, eds., Springer-Verlag, Wien.

[177] Samsonov, A.M. and Sokurinskaya, E.V. (1985). Longitudinal displacement's solitons in an inhomogeneous nonlinearly elastic rod. Leningrad, A.F. Ioffe Physico-Techn. Institute, Preprint no. 973, *in Russian.*

[178] Samsonov, A.M. and Sokurinskaya E.V. (1988a). On the excitation of a longitudinal deformation soliton in a nonlinear elastic rod. *Sov. Phys.-Techn. Phys.*, vol. 33, 8, 989-991.

[179] Samsonov, A.M. and Sokurinskaya E.V. (1988b). Nonlinear deformation waves in elastic wave guides interacting with an external medium. A.F.Ioffe Physico-Technical Institute, Preprint 1293, Leningrad.

[180] Samsonov, A.M. and Sokurinskaya, E.V. (1989). Energy exchange between nonlinear waves in elastic wave guides and external media. In: *Nonlinear Waves In Active Media*, J. Engelbrecht ed., Berlin, Springer-Verlag, 99-104.

[181] Samsonov, A.M. and Sokurinskaya, E.V. (1987). Longitudinal solitary waves in inhomogeneous nonlinearly elastic rod. *J.Appl. Mathem. Mechanics-PMM*, 51, 3, 483-488.

[182] Samsonov, A.M. et al (1996). Generation and observation of a longitudinal strain soliton in a plate. *Tech. Phys. Lett.* 22, 11, 891-893.

[183] Samsonov, A.M. and Gursky, V.V. (1999). Exact solutions to a nonlinear reaction-diffusion equation and hyperelliptic integrals inversion. *J. Phys. A*, 32, 6573-6588.

[184] Schwarz, F. (1998a). Algorithmic solution of Abel's equation. *Computing*, 60, 39-46.

[185] Schwarz, F. (1998b). Janet bases for symmetry groups. In: *Contribution to London Mathematical Society*, Lect. Notes Series 251, 221-234, B.Buchberger and F.Winkler, eds., The University Press, Cambridge.

[186] Scott, A. (1970). *Active and Nonlinear Wave Propagation in Electronics*. New York, John Wiley & Sons.

[187] Scott, A. (1999). *Nonlinear Science*. The University Press, Oxford.

[188] Shevakhov, N.S. (1977). *Sov. Phys. Acoustics*, 23, 86.

[189] Shield, R.T. (1983). Equilibrium solutions for finite elasticity. *Trans. ASME, J. Appl. Math.*, vol. 50, 1171-1180.

[190] Shutilov, V.A. (1980). *Foundations of Physics of Ultrasound*. The University Press, Leningrad, *in Russian.*

[191] Slepyan, L.I., Krylov, V.V., and Parnes, R. (1995). Solitary waves in an inextensible flexible helicoidal fiber. *Phys. Rev. Letters*, 74, 14, 2725-2728.

[192] Soerensen, M.P., Christiansen, P.L., and Lomdahl, P.S. (1984). Solitary waves in nonlinear elastic rods, I. *J.Acoust. Soc. Amer.*, 76, 871-879.

[193] Soerensen, M.P., Christiansen, P.L., Lomdahl, P.S., and Skovgaard, O. (1987). Solitary waves in nonlinear elastic rods, II. *J.Acoust. Soc. Amer.*, 81, 6, 871-879.

[194] Springer, G. (1957). *Introduction to Riemann Surfaces*. Addison-Wesley, Reading.

[195] Thurston R.N. and Brugger K. (1964). *Phys. Rev.*, 133, 1604.

[196] Tiersten, H.F. (1969). Elastic surface waves guided by thin films. *J. Appl. Phys.*, 40, 770.

[197] Toda M. (1981). *Theory of Nonlinear Lattices*. Springer-Verlag, Berlin.

[198] Toda M. and Wadati M. (1973). A soliton and two solitons in an exponential lattice and related eqs. *J. Phys. Soc Japan*, 34, 18-25.

[199] Treloar, L.R.G. (1958). *Physics of Rubber Elasticity*. The University Press, Oxford.

[200] Truesdell C. and Noll W. (1965). The nonlinear field theories of mechanics. In: *Encyclopaedia of physics*, vol. III, p. 3, Springer-Verlag, Berlin.

[201] Tvergaard, V. (1997). Studies of micromechanics of materials. *European J. Mech*, A, 1997, 5-24.

[202] Valkering, T.P. (1978). Periodic permanent waves in an anharmonic chain with nearest-neighbour interaction. *J. Phys.* A, vol. 11, 10, 1885-1897.

[203] Vlieg-Hulstman, M. and Halford, W.(1991). The Korteweg-de Vries-Burgers equation: a reconstruction of exact solutions, *Wave Motion*, 14, 267-271.

[204] West, B. (1985). *An Essay on the Importance of Being Nonlinear*. Springer-Verlag, Berlin.

[205] Whitham, G. (1974). *Linear and Nonlinear Waves*, Wiley, New York.

[206] Whittaker, E.T. and Watson G.N. (1927; reprinted 1962). *A Course of Modern Analysis*, vol.2, The University Press, Cambridge.

[207] Winkler, E. (1867). *Die Lehre von der Elastizität und Festigkeit,* Dominicus, Prague, *in German.*

[208] Wright, T. (1984). Weak shocks and steady waves in a nonlinear elastic rod or granular material. *Int. J. Solids Struct.*, 20, 9/10, 911-919.

[209] Wu, J. et al. (1987). Observation of envelope solitons in solids. *Phys. Rev. Lett.*, 59, 24, 2744-2747.

[210] Xiong, S.L. (1989). An analytic solution of Burgers-KdV equation. *Chinese Sci. Bull.*, 34, 1158-1162.

[211] Zakharov, V.E., Manakov, S.V., Novikov, S.P., and Pitaevsky, L.P. (1980). *Theory of Solitons: Inverse Scattering Transform Method.* Nauka, Moscow, *in Russian.*

[212] Zakharov, V.E. and Shabat, A.B. (1972). Exact theory of two-dimensional focusing anf one-dimensional self-modulation in non-linear media. *Soviet Physics JETP*, 34, 62-69.

[213] Zarembo, L.K. and Krasilnikov V.A. (1970). Nonlinear phenomena connected with elastic wave propagation in solids. *Soviet Physics-Uspekhi*, 102, 4, 549-586.

[214] Zhuravlev V.F. and Klimov D.M. (1988). *Applied Methods in Oscillations Theory.* Nauka, Moscow, *in Russian.*

Appendix

Explicit formuli for invariants and potential energy expressions are widely used for the statement of problems and in applications. Numerical simulations and/or large deformation problem statements can be based on the formuli derived below directly. Analytical approaches require, as a rule, further simplifications.

When torsion may be neglected, invariants of the finite strain tensor $\mathbf{C}$ depend on two displacement components (U, V) and can be written as:

$$I_1(\mathbf{C}) = \operatorname{tr}\mathbf{C} = U_x + \frac{U_x^2 + V_x^2}{2} + V_r + \frac{U_r^2 + V_r^2}{2} + \frac{V}{r} + \frac{V^2}{2r^2} \qquad \text{(A1.1)}$$

$$
\begin{aligned}
I_2(\mathbf{C}) &= \frac{1}{2}\left[(\operatorname{tr}\mathbf{C})^2 - \operatorname{tr}\mathbf{C}^2\right] = \\
&= U_x V_r - \frac{U_r^2}{4} + \frac{VU_x}{r} + \frac{VV_r}{r} + \frac{VU_r^2}{2r} + \frac{V^2 U_r^2}{4r^2} + \frac{V^2 U_x}{2r^2} + \frac{VU_x^2}{2r} + \frac{V^2 U_x^2}{4r^2} + \\
&\quad \frac{V^2 V_r}{2r^2} + \frac{V_r U_x^2}{2} + \frac{VV_r^2}{2r} + \frac{V^2 V_r^2}{4r^2} + \frac{V_r^2 U_x}{2} + \frac{V_r^2 U_x^2}{4} - \frac{V_x U_r}{2} - \frac{V_x U_r U_x}{2} - \\
&\quad \frac{V_r U_r V_x}{2} - \frac{U_x U_r V_x V_r}{2} - \frac{V_x^2}{4} + \frac{VV_x^2}{2r} + \frac{V^2 V_x^2}{4r^2} + \frac{V_x^2 U_r^2}{4} \qquad \text{(A1.2)}
\end{aligned}
$$

$$
\begin{aligned}
I_3(\mathbf{C}) &= \det(\mathbf{C}) = \\
&= -\frac{VU_r^2}{4r} - \frac{V^2 U_r^2}{8r^2} + \frac{VV_r U_x}{r} + \frac{V^2 U_x V_r}{2r^2} + \frac{VU_x^2 V_r}{2r} + \frac{V^2 U_x^2 V_r}{4r^2} + \frac{VU_x V_r^2}{2r} + \\
&\quad \frac{V^2 U_x V_r^2}{4r^2} + \frac{VU_x^2 V_r^2}{4r} + \frac{V^2 U_x^2 V_r^2}{8r^2} - \frac{VU_r V_x}{2r} - \frac{V^2 U_r V_x}{4r^2} - \frac{VU_r U_x V_x}{2r} - \\
&\quad \frac{V^2 U_r V_x U_x}{4r^2} - \frac{VU_r V_x V_r}{2r} - \frac{V^2 U_r V_r V_x}{4r^2} - \frac{VU_r V_x V_r U_x}{2r} - \frac{V^2 U_r V_r V_x U_x}{4r^2} - \\
&\quad \frac{VV_x^2}{4r} - \frac{V^2 V_x^2}{8r^2} + \frac{VU_r^2 V_x^2}{4r} + \frac{V^2 U_r^2 V_x^2}{8r^2} \qquad \text{(A1.3)}
\end{aligned}
$$

These formuli are complete and may be useful for the problem statement when elastic deformations are relatively *large*, e.g., as in case of latex, rubber-like materials, etc. After the substitution of these values of $I_k(\mathbf{C})$, the potential energy expression, containing two independent displacement components in the 5-constant theory, becomes:

$$
\begin{aligned}
\mathcal{E} \;=\; & \frac{\lambda+2\mu}{2}\left[U_x^2 + \left(\frac{V}{r}\right)^2 + V_r^2\right] + \lambda U_x V_r + \mu U_r V_x + \frac{\mu}{2}\left(U_r^2 + V_x^2\right) + \frac{\lambda}{r}V\left(U_x + V_r\right) \\
& + \frac{\lambda+2\mu+m}{2}\left(U_r^2 + V_x^2\right)\left(U_x + V_r\right) + \frac{3\lambda+6\mu+2l+4m}{6}\left[U_x^3 + V_r^3 + \left(\frac{V}{r}\right)^3\right] \\
& + \frac{\lambda+2l}{2}\left[U_x V_r\left(U_x + V_r\right) + \frac{V}{r}\left(U_x^2 + V_r^2\right) + \left(\frac{V}{r}\right)^2\left(U_x + V_r\right)\right] \\
& + \frac{2\lambda+2m-n}{4r}V\left(V_x^2 + U_r^2\right) \\
& + \frac{2l-2m+n}{r}U_x V V_r + \frac{2m-n}{2r}U_r V V_x + (\mu+m)U_r V_x\left(U_x + V_r\right) \qquad (\text{A1.4})
\end{aligned}
$$

This expression can provide the basis of correct statement of the nonlinear 2D elastic problems in polar coordinates without torsion.

The theories of rods and of thin plates provide further but different simplifications.

In rod's theory the simplest relationship $V = -\nu r U_x$ (the so-called Love hypothesis, mentioned in Chapter 2) and the plane cross section hypothesis lead to simple expressions for invariants:

$$
I_1(\mathbf{C}) = (1-2\nu)U_x + \frac{1}{2}(1+2\nu^2)U_x^2 + \frac{\nu^2 r^2}{2}U_{xx}^2; \qquad (\text{A1.5})
$$

$$
\begin{aligned}
I_2(\mathbf{C}) \;=\; & -\nu(2-\nu)U_x^2 - \nu(1-\nu+\nu^2)U_x^3 + \frac{\nu^2}{4}(2+\nu^2)U_x^4 - \frac{\nu^3 r^2}{2}U_x U_{xx}^2 - \frac{\nu^2 r^2}{4}U_{xx}^2 \\
& + \frac{\nu^4 r^2}{4}U_x^2 U_{xx}^2; \qquad (\text{A1.6})
\end{aligned}
$$

$$
I_3(\mathbf{C}) = \nu^2 U_x^3 + \frac{\nu^2}{2}(1-2\nu)U_x^4 - \frac{\nu^3}{4}(2-\nu)U_x^5 + \frac{\nu^4}{8}U_x^6 + \frac{\nu^3 r^2}{4}U_x U_{xx}^2 - \frac{\nu^4 r^2}{8}U_x^2 U_{xx}^2 \quad (\text{A1.7})
$$

Any assumptions on the values of strains were not used above, therefore terms of different order in strain component U_x were collected. It may be useful for problems in which finite and not necessarily small deformations are to be considered, e.g., for wave propagation in rubber-like materials.

When not only strains are *small* enough, $U_x \ll 1$, but also *long* waves are considered, the introduction of a small parameter:

$$
\varepsilon = \frac{U}{X} = \frac{R^2}{X^2} \qquad (\text{A1.8})
$$

yields for the Love approximation, written as:

$$V = -\nu r U_x,$$

the following values of invariants in dependence of the only component of displacement:

$$I_1(C) = \varepsilon(1 - 2\nu)U_x + \varepsilon^2(1 + 2\nu^2)U_x^2/2 + \varepsilon^3\nu^2 r^2 U_{xx}^2/2; \qquad \text{(A1.9)}$$

$$
\begin{aligned}
I_2(C) &= -\varepsilon^2(2 - \nu)\nu U_x^2 + \varepsilon^3\left[(-1 + \nu - \nu^2)\nu U_x^3 - \frac{\nu^2 r^2 U_{xx}^2}{4}\right] \\
&\quad + \varepsilon^4\left[\frac{\nu^2(2 + \nu^2)U_x^4}{4} - \frac{\nu^3 r^2 U_x U_{xx}^2}{2}\right];
\end{aligned}
\qquad \text{(A1.10)}
$$

$$I_3(C) = \varepsilon^3\nu^2 U_x^3 + \varepsilon^4\left[\frac{\nu^2(1 - 2\nu)U_x^4}{2} + \frac{\nu^3 r^2 U_x U_{xx}^2}{4}\right]. \qquad \text{(A1.11)}$$

The last terms in I_k are not used for the simplest DDE derivation (see Chapter 2) and are written here for reference. Refinement of the relationship between U and V (see Porubov and Samsonov, 1993) based on vanishing components of the Piola-Kirchhoff stress tensor on the lateral surface of a free rod yields the second term in the relationship:

$$V = -\nu r U_x - \frac{\nu^2 r^3}{2(3 - 2\nu)}U_{xxx}, \qquad \text{(A1.12)}$$

that leads to the more complicated forms of the one-component invariant expressions:

$$
\begin{aligned}
I_1(C) &= \varepsilon(1 - 2\nu)U_x + \varepsilon^2\left[\frac{1}{2}(1 + 2\nu^2)U_x^2 - \frac{2\nu^2 r^2}{(3 - 2\nu)}U_{xxx}\right] + \\
&\quad \varepsilon^3\left[\frac{1}{2}\nu^2 r^2 U_{xx}^2 + \frac{2\nu^3 r^2}{(3 - 2\nu)}U_x U_{xxx}\right] + \\
&\quad \varepsilon^4\left[\frac{5\nu^4 r^4}{4(3 - 2\nu)^2}U_{xxx}^2 + \frac{\nu^3 r^4}{2(3 - 2\nu)}U_{xx}U_{xxxx}\right];
\end{aligned}
\qquad \text{(A1.13)}
$$

$$
\begin{aligned}
I_2(C) &= \varepsilon^2(-2 + \nu)\nu U_x^2 + \\
&\quad \varepsilon^3\left[(-1 + \nu - \nu^2)\nu U_x^3 - \frac{1}{4}\nu^2 r^2 U_{xx}^2 - \frac{2\nu^2 r^2}{(3 - 2\nu)}(1 - \nu)U_x U_{xxx}\right] + \\
&\quad \varepsilon^4\left[\nu^2(2 + \nu^2)U_x^4/4 - \nu^3 r^2 U_x U_{xx}^2/2 - \frac{\nu^2 r^2}{(3 - 2\nu)}(1 - 2\nu + 3\nu^2)U_x^2 U_{xxx} + \right. \\
&\quad \left. \frac{3\nu^4 r^4}{4(3 - 2\nu)^2}U_{xxx}^2 - \frac{\nu^3 r^4}{4(3 - 2\nu)}U_{xx}U_{xxxx}\right];
\end{aligned}
\qquad \text{(A1.14)}
$$

$$I_3(C) = \varepsilon^3 \nu^2 U_x^3 + \varepsilon^4 \left[\nu^2(\frac{1}{2} - \nu)U_x^4 + \frac{1}{4}\nu^3 r^2 U_x U_{xx}^2 + \frac{\nu^3 r^2}{2(3-2\nu)}U_x^2 U_{xxx} \right]. \qquad (A1.15)$$

Each set of the invariants values provides an equation governing the nonlinear wave propagation problem under assumptions described above. The DDE for the uniform wave guide can be written as:

$$U_{tt} - c_0^2 U_{xx} = \frac{\partial}{\partial x} \left[\frac{\beta}{2\rho}U_x^2 + \frac{\nu^2 R^2}{2}(U_{xtt} - c_1^2 U_{xxx}) \right], \qquad (A1.16)$$

where the Love hypothesis was used and linear wave velocities are $c_0^2 = E/\rho, c_1^2 = \mu/\rho$, or as:

$$U_{tt} - c_0^2 U_{xx} = \frac{\partial}{\partial x} \left[\frac{\beta}{2\rho}U_x^2 - \frac{\nu(\nu-1)R^2}{2}U_{xtt} + \frac{\nu R^2 c_0^2}{2}U_{xxx} \right], \qquad (A1.17)$$

in correspondence with the refined version of the relationship $V(U_x)$.

In the 9-constant Murnaghan's theory of elasticity of compressible material under similar assumptions concerning $V(U_x)$ we have the new modification of the DDE containing more complicated nonlinearity:

$$U_{tt} - c_0^2 U_{xx} = \frac{\partial}{\partial x} \left[\frac{\beta}{2\rho}U_x^2 + \frac{\nu^2 R^2}{2}(U_{ttx} - c_1^2 U_{xxx}) + \frac{4a}{\rho}(U_x^3)_{xx} - \frac{bR^2}{2\rho}[U_{xx}^2 - \frac{1}{2}(U_x^2)_{xx}] \right], \qquad (A1.18)$$

where the coefficients are :

$$a = \frac{E(3\lambda^3 + 6\lambda^2\mu + 8\lambda\mu^2 + 4\mu^3)}{32\mu(\lambda+\mu)^2} + \frac{l}{2}(1-2\nu)^2(1+2\nu^2) + \frac{n\nu^2(1-2\nu)}{2} + $$
$$m(1+\nu)^2(1-2\nu+2\nu^2) + $$
$$\frac{\mu^4}{(\lambda+\mu)^4}\nu_1 - \frac{\lambda\mu^2(3\lambda+4\mu)}{4(\lambda+\mu)^4}\nu_2 + \nu^2(1-2\nu)\nu_3 + \nu^2(4-4\nu+\nu^2)\nu_4,$$

$$b = \frac{n\nu^3}{4} + \frac{m\nu^2}{2}(1-2\nu) + \nu^2\mu,$$

while for the refined version of $V(U_x)$, as in (A1.17), it will have the form:

$$U_{tt} - c_0^2 U_{xx} = \frac{\beta}{2\rho}(U_x^2)_x + \frac{\nu^2 R^2}{2}(U_{tt} - c_1^2 U_{xx})_{xx} + \frac{4a}{\rho}(U_x^3)_{xxx} + \frac{\nu^3 R^4}{3(3-2\nu)}U_{t4x} - $$
$$\frac{bR^2}{2\rho}(U_{xx}^2 - \frac{1}{2}(U_x^2)_{xx})_x - \frac{dR^2}{\rho}\left[U_x U_{xxx} + \frac{1}{2}(U_x^2 U_{xxx})_{xx} \right]_x - $$
$$\frac{R^4\nu^3}{3\rho(3-2\nu)}(\frac{(4\lambda+5\mu)\lambda}{2(2\lambda+3\mu)} - \mu)U_{6x} + \frac{\nu^4 R^6}{16(3-2\nu)}U_{tt6x} \qquad (A1.19)$$

where new coefficients are defined as:

$$
\begin{aligned}
a &= \frac{E(3\lambda^3 + 6\lambda^2\mu + 8\lambda\mu^2 + 4\mu^3)}{32\mu(\lambda + \mu)^2} + \frac{l}{2}(1 - 2\nu)^2(1 + 2\nu^2) + \frac{n\nu^2(1 - 2\nu)}{2} + \\
&\quad m(1 + \nu)^2(1 - 2\nu + 2\nu^2) + \\
&\quad \frac{\mu^4}{(\lambda + \mu)^4}\nu_1 - \frac{\lambda\mu^2(3\lambda + 4\mu)}{4(\lambda + \mu)^4}\nu_2 \\
&\quad + \nu^2(1 - 2\nu)\nu_3 + \nu^2(4 - 4\nu + \nu^2)\nu_4;
\end{aligned} \tag{A1.20}
$$

$$
b = \frac{n\nu^3}{4} + \frac{m\nu^2}{2}(1 - 2\nu) + \nu^2\mu; \tag{A1.21}
$$

$$
d = -\frac{\lambda^3(3\lambda + 2\mu)}{8(\lambda + \mu)^2(2\lambda + 3\mu)} - \frac{n\nu^3}{2(2\nu - 3)} + \frac{4\nu^3(1 + \nu)}{2\nu - 3}m + \frac{2\nu^2(2\nu - 1)^2}{2\nu - 3}l. \tag{A1.22}
$$

Index

Printed and bound by CPI Group (UK) Ltd, Croydon, CR0 4YY

23/10/2024

01778226-0019